Nilesha Mishra

Influence of strain on the functionality of ink-jet printed thin films and devices on flexible substrates

Schriftenreihe
des Instituts für Angewandte Materialien
Band 77

Karlsruher Institut für Technologie (KIT)
Institut für Angewandte Materialien (IAM)

Eine Übersicht aller bisher in dieser Schriftenreihe erschienenen Bände finden Sie am Ende des Buches.

Influence of strain on the functionality of ink-jet printed thin films and devices on flexible substrates

by
Nilesha Mishra

KIT Scientific Publishing

Dissertation, Karlsruher Institut für Technologie
KIT-Fakultät für Maschinenbau

Tag der mündlichen Prüfung: 12. April 2018
Referenten: Prof. Dr. rer. nat. Oliver Kraft, Prof. Dr.-Ing. Horst Hahn

Impressum

KIT Scientific Publishing

Karlsruher Institut für Technologie (KIT)
KIT Scientific Publishing
Straße am Forum 2
D-76131 Karlsruhe

KIT Scientific Publishing is a registered trademark
of Karlsruhe Institute of Technology.
Reprint using the book cover is not allowed.

www.ksp.kit.edu

Print on Demand 2019 – Gedruckt auf FSC-zertifiziertem Papier

ISSN 2192-9963
ISBN 978-3-7315-0853-3
DOI 10.5445/KSP/1000086125

Abstract

Flexible devices prepared by direct deposition of solution processable functional inks on a flexible substrate, are inexpensive, flexible, and compatible with large volume production and hence can be used for various innovative applications, which are difficult to achieve using rigid substrates. During fabrication and service, these devices are subjected to complex strains due to the use of flexible substrate, multilayer structure, encapsulation, or three-dimensional structuring, which can lead to the development of microcracks or delamination within the layers, affecting the device performance. Therefore, it is essential to understand the influence of external strain on the performance and functionality of printed thin films and thin film devices on flexible substrates.

This work is focused on fabrication of three different ink-jet printed systems on flexible substrate, characterization, and analysis of the film/ device performance in response to external strain and its correlation to specific failure mechanisms. These systems are precursor-based Indium oxide thin film transistors, Indium oxide nanoparticulate films, and Barium Strontium Titanate (BST)/ polymer nanocomposite capacitors. Micro tensile tests, in-situ synchrotron tensile tests and bending fatigue tests are performed on the printed structures to assess the device performance under uniaxial and cyclic loading conditions, respectively. Their failure mechanisms are understood by correlating their electrical or structural properties with applied strain, supported by detailed microstructural investigations.

For the precursor based printed Indium oxide transistors, the obtained field-effect mobility is 70.61 cm^2/Vs, the I_{ON} and I_{OFF} ratio is 10^7, the maximum drain current is 327µA and the devices are operational at a low voltage of around 2V. Under tensile loading, the printed transistors remain stable up to 2% strain. Beyond 2% strain the performance deteriorates. The formation of microcracks in the Indium oxide film and delamination between the film and the solid polymer electrolyte are the major factors in transistor performance degradation. The failure strain of the printed transistors is significantly affected by the quality of the printed films whereas it is insensitive to the orientation of the transistors with respect to the loading direction.

For the printed Indium oxide nanoparticulate films, in–situ synchrotron XRD is used to characterize the strain transfer between the substrate and individual printed films and small structures (e.g. Indium oxide and Au), during a single tensile test. The effect of several processing parameters on the lattice strain evolution of the individual layers is studied. The increase in lattice strain while loading of the samples, is higher for thicker samples, as compared to thinner samples. Moreover, the lattice strain evolution of the material during loading is affected by electrolyte encapsulation, introduction of Au interlayer and transistor orientation with respect to applied strain and it is sensitive to transistor geometry as lattice strain is higher for the structures.

Fully-printed Barium Strontium Titanate/polymer capacitors are fabricated at a low annealing temperature of 120°C for 1 hr and the measured capacitance and the dissipation factor is 165 pF and 0.1 at a frequency of 1 kHz, respectively. With the application of mechanical load, the capacitance at first increases and then starts decreasing gradually till the failure of the device at around 20 % tensile strain. Development of cracks in the BST layer and its subsequent propagation to the top and the bottom electrodes and in particular the surface quality of the top electrode significantly affect the device performance. It is also demonstrated that a more uniform deposition of the top Ag layer can significantly improve the stability of these devices.

Kurzfassung

Flexible elektronische Bauteile, die durch direkte Abscheidung von in flüssig verarbeitbaren funktionellen Tinten auf einem flexiblen Substrat-hergestellt werden, sind kostengünstig, nachgiebig und mit Massen-produktion kompatibel und können daher für verschiedene innovative Anwendungen verwendet werden, die mit starren Substraten nur schwer darstellbar sind. Während der Herstellung und im Betrieb sind diese Bauteile aufgrund der Verwendung eines flexiblen Substrats, Multilagenaufbau, Verkapselung oder dreidimensionaler Strukturierung, komplexen Belastungen ausgesetzt, die zur Entwicklung von Mikrorissen oder Delamination innerhalb der Schichte führen können. Und damit die Leistungsfähigkeitdes Bauteils beeinträchtigen. Daher ist es wichtig, den Einfluss von äußerer Belastung auf die Leistung und Funktionalität von gedruckten Dünnschichten und elektronischen Bauteilen zu verstehen.

Diese Arbeit konzentriert sich auf die erfolgreiche Herstellung von drei verschiedenen tintenstrahl-gedruckten Systemen auf flexiblem Substrat, deren Charakterisierung und die Analyse der Film/Bauteil-Leistung in Reaktion auf eine externe Beanspruchung und die Korrelation mit dem spezifischen Ausfallmechanismus. Diese Systeme sind precursor-basierte Indiumoxid-Dünnschichttransistoren, Indiumoxid-Nanopartikelfilme und Barium-Strontiumtitanat/Polymer-Nanokomposit-Kondensatoren. An den gedruckten Strukturen werden Mikrozugversuche, in-situ Synchrotronzugversuche und Biegeermüdungsversuche durchgeführt, um die Leistungsfähigkeit der Systeme unter einachsigen bzw. zyklischen Belastungsbedingungen zu bewerten. Ihr Versagensmechanismus wird durch die Korrelation ihrer elektrischen oder strukturellen Eigenschaften mit der angelegten Spannung untersucht, unterstützt durch detailierte mikrostrukturelle Untersuchungen.

Für die gedruckten Indiumoxid-Transistoren beträgt die erhaltene Feldeffekt-Beweglichkeit 70,61 cm^2/Vs, das I_{ON}- und I_{OFF}-Verhältnis beträgt

10^7, der maximale Drain-Strom beträgt 327 uA und die Bauelemente arbeiten bei einer sehr niedrigen Spannung von etwa 2 V. Unter Zugbelastung bleiben die gedruckten Transistoren bis zu 2% Dehnung stabil. Über 2% Dehnung verschlechtert sich die Leistung. Die Bildung von Mikrorissen in der Indiumoxidschicht und Delamination zwischen dem Film und dem festen Polymerelektrolyten sind die Hauptfaktoren für die Verschlechterung der Transistorleistung. Die Ausfalldehnung der gedruckten Transistoren wird signifikant durch die Qualität der gedruckten Filme beeinflusst, während sie unempfindlich gegenüber der Orientierung der Transistoren in Bezug auf die angelegte Spannung sind, aber empfindlich für die Transistor geometrie, da die Gitterspannung für die Strukturen höher ist.

Für die gedruckten Indiumoxid-Nanopartikelfilme wird in-situ Synchrotron Röngtenbeugung verwendet, um den Dehnungstransfer zwischen dem Substrat und den einzelnen Multi-Material gedruckten Schichten und kleinen Strukturen (z. B. Indiumoxid und Gold) während eines einzelnen Zugversuches zu messen. Die Auswirkung von mehreren Verarbeitungsparametern auf die Gitterdehnungsentwicklung der einzelnen Schichten wird untersucht. Die Zunahme der Gitterspannung während der Belastung ist bei dickeren Schichten höher als bei dünneren Schichten. Darüber hinaus wird die Gitterdehnungsentwicklung des Materials während der Belastung durch die Einkapselung durch den Elektrolyten, die Einführung einer Au-Zwischenschicht und die Transistororientierung in Bezug auf die angelegte Spannung beeinflusst, und ist aufgrund der Anwesenheit scharfer geometrischer Merkmale in Bezug auf die Transistorgeometrie empfindlich.

Die vollständig gedruckten Barium Strontium titanat/Polymer-Kondensatoren werden bei einer niedrigen Glühtemperatur von 120 °C für 1 Stunde hergestellt, und die gemessene Kapazität und der Verlustfaktor beträgt 165 pF bzw. 0,1 bei einer Frequenz von 1 kHz. Beim Anlegen der Last nimmt die Kapazität zunächst zu und beginnt dann allmählich zu sinken, bis der Kondensator bei etwa 20 % Dehnung ausfällt. Die Entwicklung von Rissen in der BST-Schicht und deren anschließende Ausbreitung in die obere und untere Elektrode und speziell die Oberflächenqualität der oberen Elektrode beeinflussen die Leistungsfähigkeit signifikant. Es wird auch gezeigt, dass eine gleichmäßigere Abscheidung der oberen Ag-Schicht die Stabilität des Bauteils signifikant verbessert.

Acknowledgments

This doctoral thesis has originated from my work at the Institute for Applied Materials (IAM-WBM) at Karlsruhe Institute of Technology from November 2012 till January 2018 under the framework of the Helmholtz Virtual Institute VI-530, 'Printed electronics based on inorganic nano-materials: From atoms to functional devices and circuits'. I would like to thank everyone who has supported me in successfully completing my Ph.D. thesis.

I would like to express my sincere gratitude for my thesis advisor Prof. Dr. rer. nat. Oliver Kraft, who as the institute leader gave me the opportunity to obtain my doctorate at the institute. I have learned a lot from his professionalism and would like to thank him for all the support and encouragement during my time at the institute.

I would like to thank my thesis supervisor, Dr. Patric A. Gruber for guiding me throughout the process. His input, ideas and our discussions has provided important contribution towards this work. This thesis would not have been possible without his constant support and encouragement.

My sincere thanks go to Prof. Dr. –Ing. Horst Hahn (INT, KIT), Dr. rer. nat. Joachim Binder (IAM-KWT, KIT), Prof. Dr. Ralph Spolenak (Laboratory for nanometallurgy, ETH Zurich, Switzerland) for creating the environment for collaborative work along with scientific advice and discussions. I gratefully acknowledge Dr. Subho Dasgupta (INT, KIT), Dr. Tessy Baby (INT, KIT) and Dr. Suresh Garlapati (INT, KIT) for their technical guidance and support. I would like to thank Mr. Morten Mikolajek (IAM-KWT, KIT) for providing the printed capacitor samples, Mr. Andreas Wyss (Laboratory for nanometallurgy, ETH Zurich, Switzerland), Dr. Alla S. Sologubenko (Laboratory for nanometallurgy, ETH Zurich, Switzerland) and Dr. Antonio Cervellino (Swiss Light Source, PSI, switzerland) for their technical support for conducting synchrotron beam line experiments at PSI, Villigen, Switzerland.

I would like to thank my group members Dr. Seol-Min Yi and Mr. Yuran Kang for their help in the laboratory and for being very supportive and helpful throughout my work. Working at IAM-WBM was a wonderful personal and cultural experience for me and for this I am grateful to all my colleagues at the institute for creating a wonderful work environment. I would also like to thank Ms. Daniela Exner and Mr. Ewald Ernst for their help in samples preparation and experimental work for the thesis.

I would like to express my deepest gratitude to my parents Subas Mishra and Suchitra Mishra, my sister Anuradha Mishra, and my sweet nephew Naman for being very supportive and understanding at every step of my life and has been a source of immense strength and happiness for me.

Finally, I would like to thank my long-time friend, colleague, and now my husband Amritesh Kumar for all his love, support, and being the pillar of strength for me.

Table of Contents

1 Introduction

Electronics in modern world have a great impact on the lifestyle of millions of people around the world. The possibility to fabricate electronic structures by printing of functional materials onto a substrate, also termed as printed electronics, has the potential to reduce the production cost significantly. Moreover, the process is simple, produces only low materials waste due to direct deposition technique, yields high throughput and is compatible with roll to roll processes to meet the demand for large scale production (Chrisey, 2000).

In addition to the adaptation of the printing technique, advancements in thin film technology and promising device architectures have led to the development of flexible electronics in recent years (Nathan et al., 2012). Flexible electronics is a collective term used for any basic electronic element such as diode, transistor, capacitor or a complete integrated circuit fabricated on a flexible polymer substrate. These electronics, integrated to a compliant substrate, can stretch, bend and twist and have specific advantages over conventional silicon electronics fabricated on rigid substrates for non-conventional and innovative applications such as smart labels, flexible displays, human wearable's, healthcare, environmental monitoring, food safety and regulation, energy conversion and storage and many more (M. Lee et al., 2012; Nathan et al., 2012; Reuss et al., 2005; Wong & Salleo, 2009). However, the printing of thin film devices onto a polymer substrate has also numerous challenges.

One of the most important challenges in the fabrication of printed flexible devices on any polymer substrate is the availability of printable inks that have the physical properties to produce high quality films with desired electrical performance, e.g. inorganic materials, which are printable, environmentally stable and have high performance. However, the processing temperature required by most of the inorganic materials are usually high and it is essential to reduce this high processing temperature in order to fit the low melting point polymer substrate (Shi & Jabarin, 2001). Flexible devices prepared at room temperature have been reported in the literature and the performances of these devices have improved steadily. But compared to the devices prepared at higher temperatures on glass or silicon substrate, the devices show inferior performance (Subho Dasgupta, Kruk, Mechau, & Hahn, 2011; Song, Jeong, et al., 2010). Therefore, a balance between processing temperature and performance is necessary to achieve optimum performance of thin film devices on flexible substrates.

The second challenge is the difficulty in developing the high quality printed films on flexible polymer substrates due to several factors such as higher surface roughness as compared to silicon or glass substrate, adhesion of the deposited material to the substrate, compliance of the substrate, proper handling of the substrate during the fabrication steps or proper alignment of different layers within the structures (Lewis, 2006). And the third and the most emerging challenge is that there is only very little information about the response of these devices to mechanical strain. Thin films are critical components in thin film devices such as transistors, capacitors, resistors, micro-electromechanical systems (MEMS) and integrated circuits and undergo significant amounts of strain depending on the type of application, during the fabrication process or while in use.

The induced strain in the structure may influence the electrical performance of the devices and finally lead to failure. Some examples of possible failure mechanisms in thin films are crack formation (C. Kim, Nogi, Suganuma, & Yamato, 2012; Nickel et al., 2014; Sim, Won, & Lee, 2012), fracture (Tsui, McKerrow, & Vlassak, 2005), fatigue failure due to cyclic loading (Gudmundson & Wikström, 2002; O. Kraft, Schwaiger, & Wellner, 2001), thin film buckling and delamination from the substrate due to poor adhesion (Sridhar, Srolovitz, & Suo, 2001) and ratcheting of encapsulated films (He, Evans, & Hutchinson, 2000; Huang, Suo, & Ma, 2002). In other words, the performance stability and reliability of such devices largely depend on the mechanical performance of thin films.

However, the mechanical properties of thin films are very different than the bulk materials due to dimensional and microstructural constraints (Arzt, 1998; Tajik & Jahed, 2006; Xiang, 2005). In addition, the use of printing technique results in a porous microstructure, where the pores act as a local necking site and hence aggravate failure. A clear understanding of the mechanical behavior of thin films and thin film devices on polymer substrate are highly essential to produce robust and reliable electronic structures for future applications. Therefore, in this work the influence of applied strain on the functionality of ink-jet printed thin films and devices on polymer substrate are studied.

This work has been performed as a part of the Helmholtz Virtual Institute VI-530: "Printed electronics based on inorganic nanomaterial: From atoms to functional devices and circuits". VI-530 is a collaboration of seven different institutes, collectively working towards the development of printed electronics based on inorganic oxides. The scope of VI-530 is primarily divided in to 3 major categories: Material Synthesis, Device Functionality, and Circuit Design. Material Synthesis deals with the development and the synthesis of printable ink-based on inorganic materials for the fabrication of thin film devices.

Device Functionality deals with the fabrication of thin film devices using the developed printable inks and characterization of the devices to optimise its performance and reliability. Circuit design deals with modelling and computing for robust circuit design using the fabricated thin film structures. This work is related to the Device Functionality and it deals with various aspects of characterization of mechanical integrity and reliability of printed thin films and devices based on inorganic metal oxides on polymer substrates. One aspect is the development of electromechanical testing techniques to correlate degradation in electrical performance of the film material with the increase in strain transferred from the substrate to the film/structure. The second aspect is performing monotonic tensile testing on the samples to determine crack onset strain and failure strain of the incorporated materials. Third aspect is performing cyclic tensile and compressive bending testing on the samples to understand the device behavior under cyclic loading and to determine the lifetime of the thin film device. All the tests are supported by microstructural characterizations of as prepared and tested films to understand the specific failure mechanism.

In Chapter 2, the topic of ink-jet printing of thin film devices on flexible substrate with a special focus on thin film transistor and metal insulator metal capacitor is discussed. It further includes discussion about current understanding of metal oxide semiconductors, mechanical behavior of thin films as well as thin film devices on flexible substrates. Different types of mechanical testing that can be performed on thin film systems to extract their mechanical properties as well as the effect of applied strain on the performance parameters of these thin film systems and the corresponding failure mechanisms are presented. In Chapter 3, experimental details of the sample preparation on flexible substrates and various testing techniques used in this work to evaluate the mechanical performance of printed thin films as well as printed thin film devices are described including set-up descriptions, testing and analysis procedures.

In Chapter 4 to 6, the results and observations from the initial electrical characterization, microstructural study and micro-mechanical testing performed on ink-jet printed Indium oxide thin film transistors on polyimide substrate, ink-jet printed Indium oxide nanoparticle films on polyimide substrate and fully ink-jet printed Barium Strontium Titanate/polymer Metal Insulator Metal (BST/PMMA MIM) Capacitor on PET substrate are presented and discussed with respect to the specific failure mechanisms.

In Chapter 7, important results are summarized and an outline of the potential future work in the field of mechanical reliability of printed devices on flexible substrates is presented.

2 Thesis background

2.1 Thin film devices

Thin films are generally defined as a layer of material having a thickness ranging from a few nanometres to several micrometres. Based on the functional behavior, thin films can be termed as active films or passive films. Active thin films are the active part of any electrical circuit and are films that respond to a change in voltage, pressure, atmospheric condition such as moisture or temperature.

Examples of active thin films are semiconductor oxides or organic thin films. Thin films, which are used in the form of coatings for applications such as decoration, thermal insulation, or anti-corrosion, are passive thin films. Examples of passive films are titania or galvanized steel. Thin film devices such as thin film solar cells, thin film transistors, capacitors and sensors can be obtained by the combination of one or more active and passive thin films (Chopra, Paulson, & Dutta, 2004; Filipovic & Selberherr, 2015; Garlapati et al., 2013). The physical properties of a thin film are strongly dependent on its microstructure and the fabrication technique used for its production.

Thin film devices fabricated by deposition of functional material/active thin films on to a flexible substrate are called flexible electronics. The advantage of using flexible substrates is that they are conformable, light, easy to produce and cost effective compared to rigid silicon or glass substrate. There are various approaches to fabricate flexible electronics, which have been already reported in literature:

1. Direct printing or solution deposition of functional materials onto a compliant substrate (Quintero et al., 2016).

2. Fabrication of an electronic device structure on a rigid substrate and with subsequent transfer to a flexible substrate (Z. T. Zhu, Menard, Hurley, Nuzzo, & Rogers, 2005).
3. Thinning of a silicon substrate to a thickness below 50 μm and then reattaching it to a flexible substrate (Zhang et al., 2009).
4. Fabrication of devices on a spin coated polymer layer on a rigid silicon substrate and subsequent peeling off the polymer layer form the wafer substrate (Mativenga, Choi, Choi, & Jang, 2011).
5. Fabrication of devices on free standing semiconductor films (Cherenack, Kattamis, Hekmatshoar, Sturm, & Wagner, 2007).

Development of flexible electronics on an industrial scale largely depends on three primary factors: development and optimization of new and existing functional materials (Forrest, 2004; Fukuda et al., 2014; M.-G. Kim, Kanatzidis, Facchetti, & Marks, 2011; S. J. Lee et al., 2016; Mahajan, Francis, & Frisbie, 2014), improvement in deposition and patterning technique, which can be integrated on a roll to roll web (Garlapati et al., 2013; Nickel et al., 2014; Sirringhaus et al., 2000; Yin, Huang, Bu, Wang, & Xiong, 2010) and assurance of long-term reliability of the structure for real life applications (Baiano, 2009; Meng, Liu, Chen, Hu, & Fan, 2010). For the development of functional materials, organic materials, inorganic materials or sometime hybrids are the kinds of materials opted to impart such flexibility, which is the most distinctive and unique characteristics of flexible electronics.

Organic materials being compatible with flexible substrates and having the ability to withstand high strains, are considered to be well suited for these kinds of applications (Berggren, Nilsson, & Robinson, 2007). But due to their lower performance and high sensitivity to atmospheric conditions, extensive research started in the area of inorganic materials especially metal oxide semiconductor materials or hybrid materials, to be used for printed devices (D.-H. Kim, Xiao, Song, Huang, & Rogers, 2010; D. H. Kim & Rogers, 2008).Metal oxide semiconductors were first reported as binary compounds such as In_2O_3, ZnO, and SnO_2. They have better electrical conductivity due to high intrinsic carrier concentration

(10^{16-21} cm^{-3}), higher optical transparency above 80 % and good environmental stability (Kwon, Lee, & Kim, 2011; Minami, 2005). Metal oxide technology has several advantages over organic technology and silicon technology such as low process complexity, easier large area processing and high electrical performance. Most of the metal oxides are n-type semiconductors and are often used for the fabrication of thin film devices such as transistors with very high-performance parameters.

In this work, thin film transistors (TFT) and metal-insulator-metal capacitors (MIM) are considered for the study as these are the building blocks of many electronic circuits. In the following sections, a basic overview of the device working principle and their important process parameters considered for performance evaluation is discussed.

2.1.1 Thin film transistor (TFT)

In 1930, Lilienfeld proposed the first concept of a field effect transistor, which is a device in which the flow of current is modulated by a transversely applied electric field (Lilienfeld Julius Edgar, 1930). Later, such devices were implemented by Shockley et al. in 1947.Thin film transistors or TFTs are three terminal devices and work on the same principle. A schematic of a thin film transistor layout is shown in figure 2.1. Source, drain, and gate electrodes are the building blocks of a thin film transistor structure. Source and drain electrodes are usually metal electrodes connected through a semiconductor channel. The gate electrode is separated from the semiconductor channel by a dielectric layer. On application of a gate potential and a voltage between source and drain electrode, flow of electrons from source electrode to the drain electrode through the channel material is established. This flow of electron from source to drain, also called as drain current, is regulated by an electrolyte/dielectric layer through the application of a gate voltage at the gate electrode.

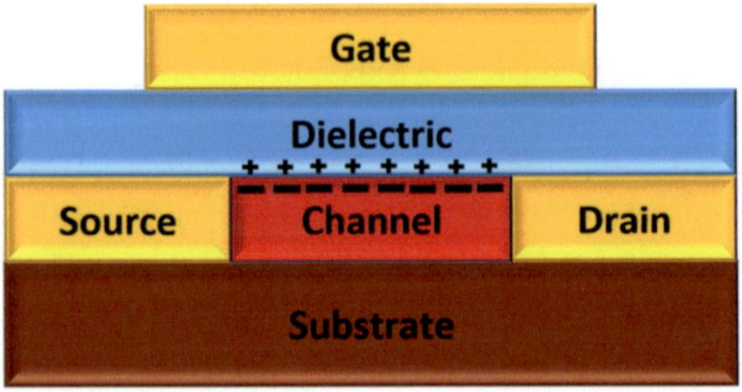

Figure 2.1: Schematic of a thin film transistor on a substrate with Source, Drain and Gate electrodes.

The first practical realization of a thin film transistor was developed by Weimer in 1962 (Weimer, 1962). Klasen and Koelmans (Klasens & Koelmans, 1964) reported the first metal oxide thin film transistor by using n-type tin oxide as the channel material and aluminium oxide as the dielectric on a glass substrate.

Later on, several metal oxide TFTs, which are fabricated through different preparation routes such as single crystal lithium doped Zinc oxide, Tin oxide, antimony doped tin oxide or Indium oxide were reported (Aoki & Sasakura, 1970; Boesen & Jacobs, 1968).

Based on the position of the gate electrode with respect to the channel layer, TFTs can be classified into three basic configurations: Top gate structure (the gate electrode is deposited on top of the channel layer), bottom gate (the gate electrode is deposited below the channel layer), and in-plane structure in which all the electrodes lie in the same plane. The top and bottom gated structures can be further classified as coplanar or staggered based on whether the source/drain electrodes are on the same or opposite side of semiconductor/dielectric interface, respectively (Tickle, 1969).

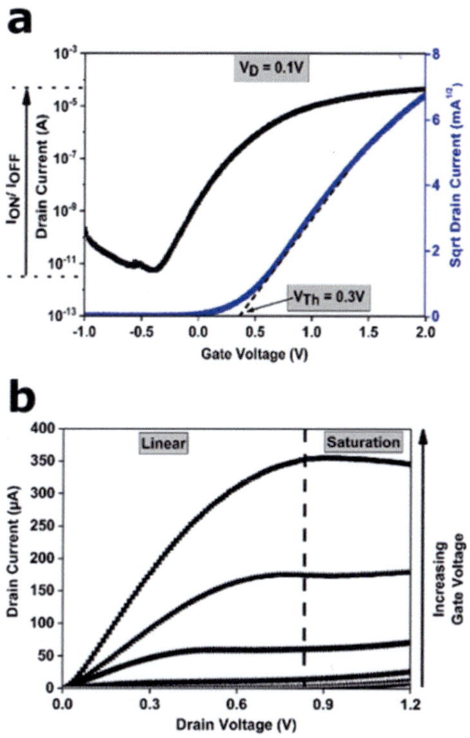

Figure 2.2: a) Transfer curve of a thin film transistor given by change in drain current with respect to gate voltage at constant drain voltage and b) Output curve of a thin film transistor showing the change in drain current with respect to drain voltage at constant gate voltage.

The performance of a transistor is given by the variation of drain current I_D with respect to the application of drain voltage V_D and gate voltage V_G. Hence, performance parameters of thin film transistors are calculated from two important characteristic curves. These are Transfer Curve I_D-V_G and Output Curve I_D-V_D as shown in figure 2.2. The Transfer Curve I_D-V_G signifies the variation of drain current with respect to gate voltage at a specific drain voltage.

The Output Curve I_D-V_D signifies the variation of drain current I_D with respect to drain voltage V_D at a defined gate voltage V_G. The applied gate voltage V_G, at which a flow of electron starts from the source to the drain electrode, is called threshold voltage V_{th} of the device. Based on the threshold voltage both curves can be divided into two specific regimes, a linear regime, and a saturation regime. If the applied drain voltage V_D is less than the difference of applied gate voltage and the device threshold voltage, the device operates in the linear regime. Beyond this value, it operates in the saturation regime.

Within the saturation regime, various process parameters for the thin film transistor such as field effect mobility, μ_{sat}, sub-threshold swing, SS, and current ratio, I_{ON}/I_{OFF}, can be calculated. The field effect mobility μ_{sat}, can be calculated as follows:

$$I_{D,sat} = \left(\frac{W}{2L}\right) C\mu_{sat}(V_G - V_T)^2 \tag{2.1}$$

where $I_{D,sat}$ is the saturated drain current, W and L are the width and length of the channel layer, respectively, and C is the double layer capacitance per unit area of the electrolyte/dielectric. The sub-threshold swing, SS, is the inverse of the sub-threshold slope and is measured in the region below the threshold voltage also known as sub-threshold region. It signifies the amount of V_G needed to change the value of I_D by one order of magnitude. I_{ON}/I_{OFF} is the ratio of maximum drain current obtained and the minimum drain current measured for the device.

2.1.2 Metal-insulator-metal capacitor

A capacitor is a device, which stores electric charge. The most simplified design of a capacitor is a metal-insulator-metal capacitor also called as MIM capacitor which has two parallel metal plates (electrodes) separated by an insulating layer (dielectric). A schematic of a parallel plate capacitor layout is shown in figure 2.3.

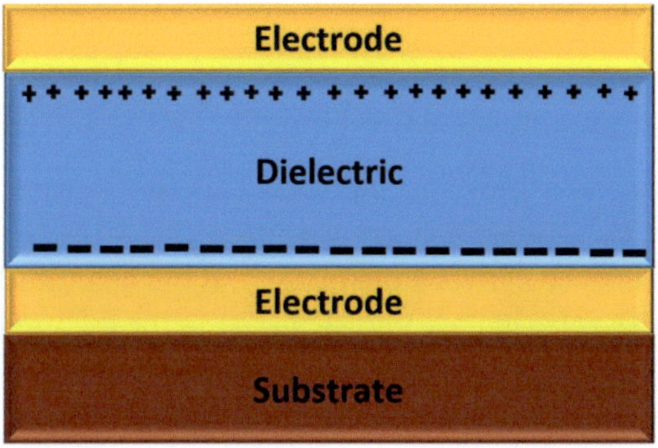

Figure 2.3: Schematic of Metal Insulator Metal Capacitor on a substrate.

When an electric field is applied across the electrodes, the dielectric gets polarized and opposite charges gets accumulated at the two electrodes, building a voltage difference between the two electrodes. This process continues until the externally applied voltage became equal to the potential in the dielectric layer. The built-up charge remains as such even after the removal of the externally applied voltage until it finds a leakage path to gradually fade. The capacity of a material to withstand this charge potential is called as its capacitance.

Ideally, a capacitor has infinite parallel resistance and the dielectric should not conduct at all. But practically there is some conductivity with in the dielectric layer resulting in a leakage current between the electrodes, which give rise to parallel resistance in the capacitor. Due to the leakage current, the capacitor starts gradually discharging. To evaluate the performance of any capacitor several process parameters are considered.

Capacitance of a parallel plate capacitor is the most important parameter. It is usually expressed in picofarad (pF) or microfarad (μF). The capacitance of a parallel plate capacitor at a specific working voltage and frequency can be calculated by equation 2.2:

$$C_P = \frac{\varepsilon_r \varepsilon_0 A}{d} \qquad (2.2)$$

where, C_P is the capacitance, ε_r is the relative permittivity of the dielectric material, ε_0 is the permittivity of air, A is the effective area of the parallel plate electrodes and d is the distance between the plates.

If two or more capacitors are connected in series, then the effective capacitance of the circuit will be considered as a single capacitor but with a larger distance between the plates and the quantitative net capacitance value of the circuit C_{Total} is calculated as the reciprocal of the total sum of the reciprocal of the individual capacitance of the capacitors:

$$C_{Total} = \frac{1}{\frac{1}{C_1}+\frac{1}{C_2}....\frac{1}{C_N}} \tag{2.3}$$

If two or more capacitors are connected in parallel, then the effective capacitance of the circuit will be considered as a single capacitor but with a larger effective plate area and C_{Total} is calculated as the sum of the individual capacitances of the capacitors:

$$C_{Total} = C_1 + C_2 \ldots\ldots + C_N \tag{2.4}$$

When an AC signal is applied the dielectric material will absorb some of the energy during the measurement. To determine the loss of the capacitor during the measurement process, the Dissipation Factor (D) is calculated. It is the ratio of energy dissipated per cycle to the stored energy per cycle. The smaller the value of D (usually <0.1) the better is the capacitor's performance.

2.2 Ink-jet printing and its challenges

Physical vapour deposition (e.g. evaporation, sputtering) and chemical vapour deposition are the most common techniques used for the fabrication of metal oxide films because of the high accuracy and high resolution that can be achieved. However, it is complex, time consuming, and involves high vacuum facilities which are very expensive and not suitable for low cost production. To overcome these restrictions, solution based deposition of metal oxides became popular due to its simplicity, low materials waste, low cost, high throughput, and compatibility with roll to roll process for large scale production even though the achievable resolution is limited. Among solution process technologies such as spin coating, spray coating, drop coating and ink-jet printing, ink-jet printing is the most promising technique for fabricating thin film structures (Yin et al., 2010).

Ink-jet printing is a popular technique for transferring digital information to a substrate. It is the process of direct non-contact deposition of functional material precisely, using layer by layer addition onto a substrate through a tiny print head nozzle. This further allows deposition of thin films of different materials whose design can be altered easily from batch to batch (Jeong & Moon, 2012; H. S. Kim, Byrne, Facchetti, & Marks, 2008; Singh, Haverinen, Dhagat, & Jabbour, 2010).

The physical and rheological properties of the ink largely define the ability of these materials to be printed. The spreading behavior of the ink is determined by the hydrodynamic properties of the ink such as Reynolds number (N_{Re}), which is the ratio of inertial to viscous forces and Weber number (N_{We}), which is the ratio of inertial and capillary forces. N_{Re} and N_{We} can be calculated as follows:

$$N_{Re} = v\alpha\rho/\eta \tag{2.5}$$

$$N_{We} = v^2\alpha\rho/\eta \tag{2.6}$$

where $\boldsymbol{v}$ is the velocity, α is the reference length, η is the viscosity, ρ is the density and γ is the surface tension. From the value of N_{Re} and N_{We}, the Ohnesorge number (1/ Z) can be obtained as follows:

$$Z = \frac{N_{Re}}{\sqrt{N_{We}}} \tag{2.7}$$

The inverse of the Ohnesorge number (Z), is a dimensionless number that determines the printability of the material. A material for which the value of Z lies between 2 and 4, is considered as printable with stable and accurate ejection of droplets, best achievable resolution and optimum performance parameters (Derby, 2010; Jang, Kim, & Moon, 2009). The ink-jet printing technique is an emerging technique and the optimization of this technique requires a multidisciplinary approach. There are various challenges which need to be considered for the future of ink-jet printing and especially for printing on flexible polymer substrates (Yin et al., 2010)

Materials availability: Even though the rapid development in the field of printable materials with high performance has directed the research towards novel large-scale cost-effective applications, the number of jettable material, although increasing, is still low. It is because the usage of printable functional material depends on various factors such as functional performance, print quality, compatibility and adhesion to the substrate, jetting characteristics, ease of manufacturing and reliability (Magdassi, 2009). To get these materials to meet the above requirements is a challenge.

Moreover, noticeable progress has been reported for N-type semiconductor materials, P-type semiconductor materials can hardly match to their N-type counterparts in terms of performance especially in the case of printed oxide semiconductors (Anthony, Facchetti, Heeney, Marder, & Zhan, 2010; Herlogsson, Coelle, Tierney, Crispin, & Berggren, 2010). As the realization of compact low power circuits largely depends on the development of complementary circuits based on both P-type and N-type semiconductor materials, the development of solution processed P-type devices with higher performance is a future challenge. In addition, for the adaption to flexible substrates, low temperature processing throughout the device fabrication process is needed.

Dimensional instability: Printing of materials on a substrate is even more challenging when done on a flexible substrate. This is due to increase in the complexity of the fabrication process and circuit integration due to the dimensional instability of the substrate.

Film quality: The quality of the film obtained through printing technique is often porous and non-homogeneous, which results in poor quality of the film as compared to thin films obtained through vacuum deposition techniques. In addition to that, due to higher surface roughness of the flexible substrates as compared to conventional silicon or glass substrates it is very challenging to get good quality printed films on flexible substrates.

Induced strain: Strain may be induced to the printed flexible devices during the device fabrication process due to bending or stretching of the substrate or while in use. These induced strains can be very high and have a significant impact on the device performance and functionality.

2.3 Mechanical properties of thin films and devices

With the progress of research and development in microelectronics, thin films became a major part of integrated circuits (Anderson, 1972). Thin film devices usually incorporate complicated structures, layering of different materials and various small integrated features to achieve a desired characteristic or a property. For a reliable operation of these devices, it is very essential that its individual layers should be able to retain their integrity and electrical properties for a long period of time. Therefore, the necessity to understand the mechanical performance of these thin films is crucial to predict the device performance and reliability during its life time.

The mechanical properties of thin films are very different than the bulk materials due to microstructural and geometrical constraints (Arzt, 1998). In the bulk metals the mechanical strength is largely dominated by dislocation interaction and pileup during the plastic deformation of the material. However, in thin films as the spacing between the dislocations are comparable to the microstructural feature size of the material (Hall, 1951; Petch N J, 1953) (Brenner, 1956), the dislocation network is constricted to a small volume. As a result, the flow stress of metal thin films is significantly higher than that of the bulk materials and further increases with decreasing film thickness (Nix, 1989). W. D. Nix in his work 'Mechanical Properties of Thin Films' (Nix, 1989) discussed the critical thickness theory for thin films on substrate based on the channelling dislocation model suggested by Matthews and Blakeslee for epitaxial semiconductor thin films (Matthews, 1974).

In his theory, Nix has predicted that the strength of thin metal films is inversely related to its thickness. Later Nix's theory was confirmed by several experimental findings for films below 1 µm film thickness (Griffin, Brotzen, & Dunn, 1992; O. Kraft & Volkert, 2001). In addition to film thickness, grain size (Doerner, Gardner, & Nix, 1986) and surface texture (Sanchez & Arzt, 1992) of the film also play an important role in the deformation behavior. The dependence on size effects observed in thin films and small size structures has been elaborately discussed in the work of Kraft et al (Oliver Kraft, Gruber, Mönig, & Weygand, 2010). In very thin films (film thickness <100 nm) the strength of the material during plastic deformation is largely dominated by partial dislocations as for such thin films the nucleation of partial dislocations is easier compared to full dislocations (M. Chen, 2003).

Based on the fabrication route, the processing temperature, the substrate used and its application different types of stress can be generated in a thin film system (Evans & Hutchinson, 1995)(Hutchinson & Suo, 1991). These stresses can be primarily categorized into two important categories: intrinsic stresses and extrinsic stresses. Intrinsic stresses are generated within the thin film material during the process of material deposition or due to difference in thermal expansion coefficient of the film material and the substrate. Extrinsic stresses mostly include stresses, which are generated within a thin film material during its use. Increasing stresses in the film may result in thin film failure in the form of cracking, delamination, and deformation of the film.

Thin metal films, either freestanding or substrate supported, can rupture sometimes at a very small tensile strain below 1% (Chiu, Leu, & Ho, 1994; Erdem Alaca, Saif, & Sehitoglu, 2002) or in some cases, can withstand strains beyond 50% (P. Gruber et al., 2004; Lu, Wang, Suo, & Vlassak, 2007; Macionczyk & Brückner, 1999). This contradictory findings by different groups were addressed and explained by the group of Suo et al (Zhigang Suo, Vlassak, & Wagner, 2005). According to them, the failure mechanism in thin films depends on several factors such as whether the film is free-standing, or polymer supported, the degree of film-substrate adhesion and their microstructure.

When a freestanding metal film is subjected to monotonous tensile strain, the film at first deforms plastically up to certain amount of strain and then the process is followed by a preferential thinning of the film (also called necking) as shown in figure 2.4a. This localizes the deformation in that area leading to final fracture of the film (Teng Li & Suo, 2006). However, when the film is supported by a polymer substrate as shown in figure 2.4b, the strain localization is constrained by the substrate underneath and hence the material can sustain larger strain due to a uniform deformation over a larger area (P. Gruber et al., 2004; T. Li et al., 2005; Teng Li & Suo, 2006; Xiang, Li, Suo, & Vlassak, 2005).

Another prominent failure mechanism in thin metal films on polymer substrates is grain boundary cracking/rupture by cleavage formation as shown in figure 2.4d and 2.4c, particularly seen in brittle material. It is the process of breaking of the array of atomic bonds at the grain boundary when the applied tensile strain reaches the critical failure strain of the material. In this case the metal film fails with negligible elongation in the tensile direction and the support of polymer substrate does not have any significant effect on increasing the rupture strain of the material.

Adhesion plays a key role in determining the interface quality of the film/substrate system. A strong interface can help in increasing the strain tolerance capacity of the film whereas a weak interface between the metal film and the polymer results in interfacial cracks or de-bonding/ delamination of the film from the substrate as shown in figure 2.4c. Under small amount of tension, a weak interfacial adhesion can result in delamination of the film from the substrate.

The film becomes free-standing and free to form a neck or grain boundary cracking. On the other hand, a metal film with a strong interfacial adhesion between the film and the substrate can withstand a strain of up to 50 % or beyond. This has been verified experimentally by conducting tensile tests on Cu thin films in which weakly bonded Cu films on polyamide substrate failed at around 2 % tensile strain whereas well bonded Cu films can sustain a tensile strain of 50 % without fracture (Lu et al., 2007; Xiang et al., 2005).

Along with the interfacial quality, the microstructure of the film material also plays an important role in the deformation mechanism of metal films on polymer substrate. Ink-jet printing process is used in this study for the device fabrication process and the printed films inherently have a nanoporous microstructure. Therefore, it is also necessary to understand the failure mechanism of films with porous microstructure. A metal film on polymer substrate with nanoporous microstructure has a lower failure strain when subjected to tensile strain as compared to a dense metal film. This is because the nanopores as shown in figure 2.4f act as local necks and hence initiate crack nucleation in the film.

A metal film adhering to a polymer substrate can experience interplay of one or more of the mechanisms discussed above and fail in several ways. One of the ways is thin film delamination from the substrate due to propagation of a crack at the film/substrate interface, or, if it is a brittle film, the crack can even leave the interface and propagate into the substrate. The mechanism of thin film delamination depends on several factors such as in-build residual stress in the film, presence of pre-existing defects and interfacial strength. Another way, mostly seen in brittle materials is crack nucleation and propagation perpendicular to the interface of the film/substrate system.

However, if the film is strained in tension, a series of parallel through-thickness cracks perpendicular to the loading direction will develop. The development of cracks partially relieves the stresses in the film. Within the crack the stress is zero, whereas at some distance from the crack path the stress is not relaxed due to ongoing strain transfer from the substrate into the film. With further straining, additional parallel cracks form, resulting in decreasing crack spacing. The crack spacing depends on the film thickness, the energy release rate of the parallel cracks and the elastic mismatch between the film and the substrate (Beuth, 1992).

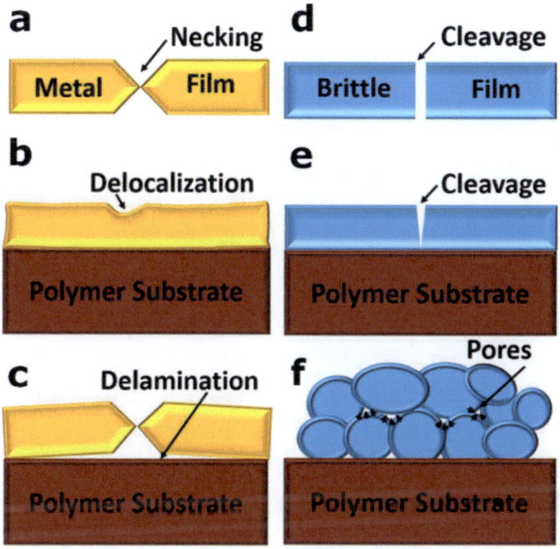

Figure 2.4: Schematic of failure mechanics of a) free standing metal thin fil
metal thin film on a polymer substrate with strong adhesion and weak ad
tively, d) free standing brittle thin film e) brittle thin film on a polymer
nanoporous thin film on a polymer substrate.

A detailed review of the mechanical characterizatio
tors on flexible substrates has been reported by
(Petti et al., 2016) titled 'Metal oxide thin fil
electronics'. Fabrication of thin film tran

oxides precursors or nanoparticles such as Indium oxide, Indium tin oxide, tin oxide, Zinc oxide (Avis, Hwang, & Jang, 2014; Baby et al., 2015; Garlapati et al., 2013, 2015; Song et al., 2009) are currently in demand because of the inherent materials quality such as high transparency, uniformity, device performance (due to high intrinsic mobility) and environmental stability as compared to organic or Si based TFTs (Yin et al., 2010). However, the mechanical characterization of these high performing devices are still premature (Subho Dasgupta, Kruk, Mechau, & Hahn, 2011; Hwan Hwang et al., 2013; Ju et al., 2007; M.-G. Kim et al., 2011; Y.-H. Kim et al., 2012; C. Y. Lee et al., 2010; Lin et al., 2013; J. H. Park et al., 2013; Seo et al., 2013).

In-general, inorganic semiconductor materials can withstand less than 1% of tensile strain (M.-J. Park et al., 2015), but several ways have been suggested in the literature to improve the stretchability of the metal-oxide thin film transistors by modification in the device design or in the fabrication route. Park et al (S.-I. Park et al., 2008) studied both experimentally and theoretically the bending of inorganic electronic materials on plastic substrates. They reported that thin layers of silicon weakly bonded to plastic substrates exhibit three different failure modes at sufficiently high bending strains: cracking, slipping, and delamination, depending on the thickness of the silicon layer. These failure modes are controlled by surface strain, interfacial shear stress, and interfacial normal stress, respectively. Although their work was primarily focused on single crystalline silicon ribbon deposited on plastic substrate, but their reported guideline is useful for designing flexible inorganic system on plastic substrates.

Khang D Y et al.(Khang, 2006) also suggested a unique approach to [illegible]nprove the mechanical performance of inorganic materials by deposit[illegible] single crystalline Si in a wave like structure on an elastomeric sub[illegible]te. By re-adjusting the amplitude and distance between the waves [illegible]naterial can withstand comparatively higher strain in this kind of [illegible]ecture. Similar approaches were also reported by few others with a [illegible]quantitative approach towards the relation between the strain accommodation and change in wave parameters (Jiang et al., 2007; Lacour, Chan, Wagner, Li, & Suo, n.d.). There are also reported results to control the wave dimensions at micro and nanoscale to degenerate structures with defined geometry and shape as well as improved mechanical performance by regulating their buckling wave's orientation, amplitude and distance (Jiang, Sun, Rogers, & Huang, 2008; Wagner et al., 2004).

Li and Suo (Teng Li & Suo, 2006) have shown that when a stiff material is specifically patterned on a compliant substrate in such a way that the brittle film can take a very small amount of strain by twisting out of plane when a large strain is applied to the substrate. Besides, the mechanical reliability or performance of inorganic thin films can be further improved by positioning the active film of the device within the neutral plane of a bending structure to reduce the amount of actual strain to the inorganic film (Sekitani et al., 2005; Z. Suo, Ma, Gleskova, & Wagner, 1999). The development of nanocomposite films in particular as a combination of ceramic material and polymer makes it feasible to produce films, which combine properties from both the materials.

That means ceramic/polymer composite films have high permittivity and low dielectric loss inherited from the ceramic material and flexibility and low processing temperature inherited from the polymer material (Gevorgian, 2009; Kohler et al., 2013; L. Zhu & Wang, 2012). In combination with printing technology, nanocomposite dispersions can be used to fabricate flexible electronics (Mahajan et al., 2014; Nickel et al., 2014; Sirringhaus et al., 2000).

There are some reported ink-jet printed MIM capacitors, which use additional vacuum processes to complete the fabrication steps in addition to printing technique (Sriprachuabwong, Srichan, Lomas, & Tuantranont, 2010). Whereas some are also reported to be fabricated on non-flexible substrates (Liu, Cui, & Varahramyan, 2003) or at higher temperature (Lim et al., 2012). Although a lot of work has been done towards development of flexible electronics, most of the work is primarily focused on the development of new materials, optimization of the structure design and the process parameters (Cantarella et al., 2015; H. Chen, Rim, Jiang, & Yang, 2015; Chang-Ho Choi et al., 2015; Fukuda et al., 2014; Hwan Hwang et al., 2013; M.-G. Kim et al., 2011; Ko et al., 2007; Mahajan et al., 2014; Song, Jeong, et al., 2010; Woo, Cherenack, Tröster, & Spolenak, 2010). However, for successful implementation of this technology, a better understanding of the mechanical failure behavior of the flexible devices will be necessary for developing products which have stable mechanical and functional performance and reliability during its life time.

2.4 Mechanical testing of thin films and devices

As has been discussed so far it is difficult to extract mechanical properties of thin films by simply applying the general mechanical testing methods used for bulk materials due to small material volume available for the testing procedure, the difficulties in handling of the small specimen and the necessity to measure force and displacement of very low magnitude. Moreover, unlike for bulk materials, there is no standard procedure or methodology to follow for mechanical testing of thin films.

However, several test methodologies have been developed and used to measure the mechanical properties of thin films both for free standing as well as polymer supported thin films or small structures. These are, tensile testing, micro-beam bending, bending fatigue testing, bulge testing, depth sensing indentation, curvature measurement, thermal cycling and micro compression testing. Elastic, plastic and time dependent properties of thin films such as Young's modulus, flow stress, fracture strength, fatigue resistance and creep can be determined by these techniques but each of them has their own advantages and disadvantages. Critical overviews of these techniques have been reported in the following references (Griffin et al., 1992; O. Kraft & Volkert, 2001; Nix, 1989).

One of the most common testing techniques used to extract thin film properties is thermal cycling. Thin films and small structures adhering to a substrate develop residual stresses when subjected to repeated changes in temperature. This is due to the difference in thermal expansion coefficient of both the materials, as the film changes its dimension with respect to the substrate and hence develops residual stresses.

Residual stresses result in a curvature of the film/substrate system, which can be measured to determine the stress within the film (Flinn, Gardner, & Nix, 1987; Keller, Baker, & Arzt, 1998). The scope of the thermal cycling method to obtain the mechanical properties of thin films is limited because the maximum strain that can be achieved during a thermal cycle is only around 1 % and the stress and temperature for the thermal cycling process cannot be varied independently.

Depth-sensing hardness measurements, which measure the indentation depth as a function of the applied load, are well suited for thin film systems. Hardness, which is a combination of elastic and plastic properties, can be calculated by measuring the indention depth or contact area at a given load and correlate them (Oliver & Pharr, 1992). However, due to complexity in the analysis of hardness measurement, its scope is also limited. Tensile testing of either free standing or substrate supported thin films can be performed to obtain their mechanical properties.

Tensile testing of free standing film is useful in giving insight into the material's intrinsic property and is easier to interpret as compared to substrate supported films as the data obtained is not mixed, whereas the latter case the obtained data is a combination of substrate and film and hence the signal of the films is very small. However, handling of the sample is easier in the latter case.

For the substrate supported film, the evaluation of tensile data to obtain film properties is done in several ways. One of them is to perform similar tensile test on bare substrate and film/substrate system and to calculate the film properties from the difference between the two stress-strain curves (Kang, Ho, Knipe, & Tregilgas, 1996; Macionczyk & Brückner,

1999). One of the drawbacks of this method is that the effect of biaxial deformation due to the mismatch in Poisson ratio of the film and the substrate is ignored. Another way to calculate the intrinsic property of the film from the substrate supported film is in-situ strain measurements by X-ray diffraction during the tensile test (Noyan & Sheikh, 1993).

X-ray strain measurement in combination with tensile tests has been studied on both metallic substrate such as Ni or Al (Noyan & Sheikh, 1993; Renault et al., 1998) and polymer substrates. But the use of polyimide substrate such as DuPont™ Kapton® has an added advantage of excellent surface quality and homogenous deformation for even more than 5% strain (O. Kraft, Hommel, & Arzt, 2000). Further use of elastic substrate, also allows to study of deformation behavior of the thin film under tension and compression as the plastically deformed film goes under compression while unloading the system (Hommel, Kraft, & Arzt, 1999). Frank et al. (Frank, Handge, Olliges, & Spolenak, 2009) monitored the evolution of the biaxial stress state of Ta films during fragmentation and buckling using synchrotron XRD (sXRD).

Using the sXRD technique, the size effect on yield strength and strain hardening of thin Cu films was studied by Gruber et al. (P. A. Gruber et al., 2008). Gruber et al. showed that for a copper film the yield strength increases with decreasing film thickness and for a film thickness of 31 nm its yield stress reaches its maximum of 623 MPa. Lohmiller et al. showed that the asymmetry of the peak can be used to differentiate deformation mechanisms in nanocrystalline films (Jochen Lohmiller, Baumbusch, Kraft, & Gruber, 2013; Jochen Lohmiller, Kobler, Spolenak, & Gruber, 2013). Besides a systematic study has been conducted to investigate the cracking behavior of Cu films with Ta interlayer on polyimide with variable thickness (P. A. Gruber, Arzt, & Spolenak, 2009). Here, the synchrotron-based tensile testing technique has been combined with in situ tensile tests in a scanning electron microscope. It is found, that the fracture toughness of the film system increases with increase in the thickness of the metal film and decreases with the thickness of the interlayer film.

For the overall performance evaluation of thin film devices especially designed for flexible applications, characterization of only its electrical parameters is not sufficient. Rather a systematic evaluation of mechanical properties such as induced strain in the device, maximum strain resistance, bendability, influence of cyclic loading, effect of strain on electrical performance or strain transfer within the layers are some of the important issues that need to be considered. Therefore, electro-mechanical characterization of the flexible thin film devices is crucial in order to understand the impact of the film deformation and fracture on the functionality and reliability of the printed devices.

Fragmentation testing is one of the common testing methods employed to observe the mechanical behavior of metal films on polymer substrate. In this method film/substrate system is observed under a microscope during the uniaxial tensile straining of the sample. Cordill et. al (Cordill, Glushko, Kreith, Marx, & Kirchlechner, 2015) conducted an electro-mechanical study on Cu films on PET substrates by combining, measurement of film resistance by in-situ 4-point probe, imaging of the film microstructure during straining by in-situ atomic force microscopy and in-situ synchrotron tensile tests. In-situ AFM imaging helped in quantifying the crack spacing developed during the tensile test and this is then compared to the change in film resistance measured though 4-point probe to correlate the film deformation mechanism with the change in film resistance during straining.

XRD stress-strain behavior measured for the same samples correlated well to the observed failure mechanism. Lu et al. (Lu et al., 2007) conducted experiments on Cu films deposited on bare polyimide substrates and on polyimide substrates with an addition of Cr interlayer. It is shown that the Cu films fail due to formation of micro-cracks in the films as a result of the coevolution of film debonding from the substrate and of strain localization. Also, addition of Cr interlayer improves the film/ substrate adhesion, thereby retard strain localization, and cracking of the Cu films.

Among the various testing techniques, bending testing is the most common type of testing method used so far for the characterization for printed device/structure. Several group has shown the effect of single bending test on flexible TFTs at a certain bending radius (Petti et al., 2015; Rim et al., 2014; Yeo et al., 2013; X. Yu et al., 2015) and under cyclic bending loading conditions of few cycles (S. Kim, Won, Sim, Park, & Lee, 2013; J. H. Park, Oh, Han, Lee, & Baik, 2015; Thiemann et al., 2014).

For example Jun et al (Jun, Park, Cho, & Kim, 2009) have shown the effect of bending on the field effect mobility of the TFTs under investigation. They have shown that the application of tensile load increases the particle to particle distance and hence decreases the conductivity of the nanoparticle network. Thiemann and group (Thiemann et al., 2014) have shown that application of tensile bending radii of around 8.5 mm leads to crack formation and results in early device failure in the case ZnO TFTs whereas the devices were fully operational up to compressive bending radii of 1.1 mm. Precursor based TFTs on flexible substrate are reported to fail with application of strain mainly due to formation of crack in the most brittle layer of the device i.e. strain induced breakdown of the gate dielectrics, formation of crack in the source, drain or gate contacts (Jeong et al., 2013; Song, Noh, et al., 2010). Kim et. al. (B. Kim et al., 2014) have shown that it is possible to get a highly conductive and reliable nanoporous Ag film by optimizing the nanoporous microstructure.

On the one hand, the nanoporous microstructure of the films imparts good electrical conductivity due to particle to particle connections. In addition to that, it also reduces the plastic deformation during the sample is subjected to cyclic loading and therefore reduces the formation of fatigue damage. This can be a guideline for producing high quality printed structure with long term reliability. In other words, there are some previously reported works correlating the effect of mechanical bending and uniaxial straining on the performance of organic thin film transistors and individual transistor components (Fukuda et al., 2014; Jochen Lohmiller et al., 2014; Sekitani, Zschieschang, Klauk, & Someya, 2010;

Takahashi et al., 2010) but not much investigation has been done on printed inorganic materials.

In addition to flexible printed transistors, development of flexible printed capacitors has also been explored recently due to high stability and low nozzle clogging of dielectric inks during the printing process (Minemawari et al., 2011) One of the promising ways to improve the properties of printed capacitors further is the use of nanocomposite films, which makes it feasible to produce devices having optimum combination of the properties from both types of materials used in the nanocomposite. Most of the previously reported work have been focused on the development of ceramic-polymer composite dispersions having good printing ability and stability (Kohler et al., 2013), (L. Zhu & Wang, 2012), (Singh et al., 2010), (Friederich et al., 2014). Vescio et al (Vescio et al., 2016) have reported a study on fabrication of fully ink-jet printed HfO_2 based MIM capacitors on flexible substrate.

To enable the integration of thin film electronics into any kind of conformal or soft surfaces, evaluation of the stretchability of the individual layers as well as of the whole devices is necessary. Based on the type of intended applications the requirement of the mechanical performance of these thin film devices is different and correspondingly the types of mechanical tests used to transfer the induced strain from substrate to the printed structures are also different. For example, for application like rollable displays, RFIDs, patches etc. a bendability of few millimetre would be enough, whereas for wearable textiles a bendability of sub millimetre range would be required (Zysset, Munzenrieder, Kinkeldei, Cherenack, & Troster, 2012). These kinds of applications do not require a high level of stretchability in the electronics. However, microelectronic devices targeted for applications such as biomedical implants or artificial skin etc. should have a mechanical flexibility close to human skin i.e. they should be bendable as well as stretchable up to 70 % (Münzenrieder et al., 2015). Moreover, out of different modes of deformation, stretchability can easily induce a large tensile strain with in the sample and so it is the most challenging case.

2.5 Aims of the study

Due to its high intrinsic mobility and better environmental stability, metal oxide semiconductors can significantly improve the performance of thin film devices. The use of ink-jet printing technique to fabricate metal oxide devices makes the process simple, low cost, and feasible to different types of substrates.

In general, printed thin film transistors can be fabricated by two different fabrication routes i.e. precursor route and nanoparticulate route. The use of precursor route in the transistor fabrication process results in high quality of metal oxide film but involves the use of high temperatures during the fabrication, which inhibits the use of polymer substrates. While the use of nanoparticulate route, can significantly reduce the processing temperature for the fabrication process but results in a porous microstructure. The electrical properties of the flexible devices prepared through precursor-based route is superior to the devices prepared through nanoparticulate route, however, the mechanical reliability of the devices prepared through these two different routes have not been examined in detail. Therefore, thin film devices prepared from both the fabrication route have been studied in this work, to establish the best route for the preparation of these flexible printed devices, with respect to the electrical as well as mechanical properties.

Over the last decade, a lot of research has been directed towards improving the electrical properties of flexible thin film devices, while the mechanical stability and reliability of these devices has been largely ignored. Mechanical reliability of these flexible devices is crucial because the use of these flexible printed devices in commercial applications require a minimum standard for mechanical stability and reliability. The mechanical performance of these flexible printed devices depends on the type of substrate used, microstructure of the film, adhesion between film and substrate and the intended application.

There are some previously reported works correlating the effect of mechanical bending and uniaxial straining on the performance of organic thin film transistors and individual transistor components as discussed before. However, the effect of external strain on ink-jet printed oxide thin film transistors have not been widely explored so far. Therefore, it is essential to study the effect of mechanical strain on the performance of the individual components and the flexible printed devices in order to understand the failure mechanisms of flexible printed devices to predict its mechanical stability and reliability during its lifetime.

In addition to flexible printed transistors, development of flexible printed capacitors has also been explored recently due to high stability and low nozzle clogging of dielectric inks during the printing process by using nanocomposite films, which makes it feasible to produce devices having optimum combination of the properties from both types of materials used in the nanocomposite. Most of the previously reported works have been focused on the development of ceramic-polymer composite dispersions having good printing ability. However, again, not much has been reported about the mechanical performance and reliability of these printed nanocomposite capacitors on flexible substrates.

Considering the factors described above, the aims of this study are the following: To study the influence of strain on the functionality and reliability of printed metal oxide based flexible devices through electromechanical characterisation.

1. To understand the importance of mechanical stability of the printed structures having different microstructure, understanding their failure mechanisms and to optimize the mechanical reliability of such components by varying the device design, the fabrication process and the microstructure of the incorporated materials.
2. To understand the strain transfer between the flexible substrate and the metal oxide nanoparticulate films printed on the substrate using a synchrotron-based in situ testing technique.

3 Experimental details

3.1 Sample preparation

The fabrication process of printed films and devices is explained in this section. The first section includes details about the fabrication process of precursor-based Indium oxide thin film transistors (TFTs) on polyimide (PI) substrate, followed by fabrication of Indium oxide nanoparticulate (np) films also on PI substrate and finally fully ink-jet printed Barium Strontium Titanate (BST)/Polymethylmethacrylate composite metal insulator metal capacitor on Polyethylene terephthalate (PET) substrate.

Printing of functional semiconductor oxides to fabricate thin film transistor can be done by using either a chemical precursor of the semiconductor oxide or by dispersion of semiconductor nanoparticles. For the precursor route metal salts such as Indium acetate, Indium chloride, or Indium nitrate are dissolved in a suitable solvent and allowed to decompose into respective oxides to get the ink with the desired properties. The ink is used subsequently for the printing process.

The printed precursor salt usually needs a high temperature annealing step to decompose into the respective oxide. The printed oxide film obtained through precursor route has a good quality and superior electrical performance (C.-H. Choi, Lin, Cheng, & Chang, 2015). However, requirement of high processing/annealing temperature makes it not suitable for many flexible substrates such as PET.

Several methods has been reported in literature such as sol-gel method (Banger et al., 2011), UV curing , photonic curing (Y.-H. Kim et al., 2012), or combustion synthesis (M.-G. Kim et al., 2011) to reduce the processing temperature of these metal oxides but still it's a major challenge to get high quality printed films on compliant substrates.

For the nanoparticulate route, metal nanoparticles with a suitable size distribution with respect to the ink-jet printer nozzle diameter (e.g. 20-50 nm particles for nozzle diameter of 20 µm) are dispersed in a solvent and a stabilizing agent. The stabilizing agent prevents the coagulation of the nanoparticles within the dispersion. After printing of the nanoparticulate ink the solvent and the stabilizing agent evaporates resulting in formation of nanoparticulate network which can form a conducting path. Since no thermal annealing step is needed in this case, the process is suitable for all kinds of substrates. However the quality of the printed film is usually inferior to that of the precursor route as the nanoparticulate film often has poor particle to particle connectivity as well as high porosity (Subho Dasgupta et al., 2011).

3.1.1 Indium oxide TFTs by precursor route

Polyimide was chosen as substrate for the fabrication of printed In_2O_3 TFTs using the precursor route. This is because polyimide in general has a small coefficient of thermal expansion of 12 $x10^{-6}$/K, high glass transition temperature (T_g) of 355°C and a surface roughness in the nanometer regime (Kirstein, 2013). The substrate was cleaned by ethanol and dry N_2 gas followed by oxygen plasma cleaning of 5 min at 100 W in a plasma cleaning machine. The cleaning process is essential to remove dust particles from the substrate surface and to improve the quality of the surface before deposition of electrodes.

The chosen transistor design is a simple in plane geometry with a fixed channel length (L) of 100 µm and channel width (W) of around 100 µm. The dimension of the source (S), the drain (D) and the gate contact pad (G) are around 300 x 300, 300 x 300, 500 x 500 µm^2 respectively. The gap between the S, D and G electrodes are 100 µm. The length and the width of the S and D electrodes are 10 µm and 600 µm respectively. Figure 3.1 shows the detailed geometry of the fabricated transistors. Printed thin film transistors on rigid substrate with similar in-plane geometry has been reported earlier (Subho Dasgupta et al., 2011; Garlapati et al.,

2013). The in-plane geometry for the transistor fabrication is chosen in this study because of two reasons. One is to complete the fabrication step for S, D and G in one single step and the second reason is the electrolyte which is used as the dielectric material which cannot withstand ultra-high vacuum. It is because of the decrease in ionic conductivity of the electrolyte due to evaporation of the trapped solvent molecules under vacuum condition.

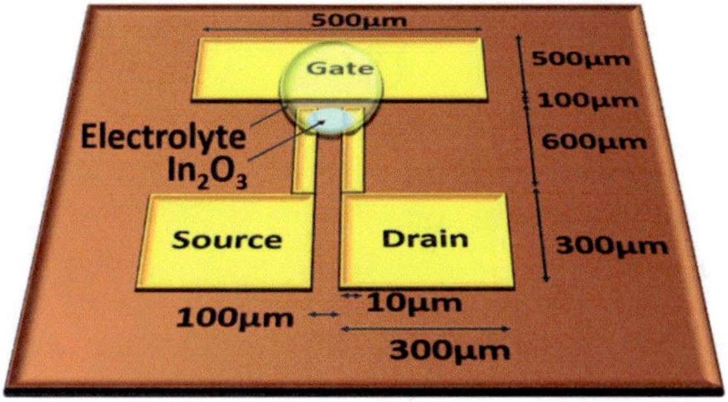

Figure 3.1: Schematic of an in-plane thin film transistor showing, Source (S), Drain (D) and Gate (G) electrodes with printed semiconductor channel and electrolyte.

The electrodes were patterned by thermal vapour deposition (Leybold Univex 350G) of Cr and Au having a thickness of 10 and 40 nm, respectively, and using a metal mask (Laser job GmbH, Germany). The total thickness of the metal mask is 0.2 mm. The patterned substrate was further heated on a hot plate at 200°C for 1 hr. to remove any unwanted organic residues before printing of the channel layer.

The concentration of the precursor ions determines the final device performance of the printed transistor. A very high concentration of the precursor results in a highly conducting film which will result in a transistor having a high off current, which means the transistor, is difficult to switch off. Whereas low loading of the precursor results in film with not enough electrical conductivity to show the desired transistor behavior. Therefore, optimization of the precursor concentration in the precursor solution is highly essential. Also, quality of the printed layer is not only important for better electrical performance, but it is also important in terms of mechanical response of the printed devices when subjected to external strain. To understand the effect of film quality on the mechanical behavior of printed TFTs, the Indium oxide transistors were prepared as two different systems.

In the first system, an optimized precursor solution was prepared with 0.05 M of Indium Nitrate Hydrate ($In(NO_3)_3 . xH_2O$) from Sigma Aldrich in 10 ml of water and ethanol in equal proportion. After mixing the solution thoroughly, the solution was filtered using a 0.2 µm polyvinylidene fluoride (PVDF) membrane filter. Filtration process is done to prevent clogging of the print head nozzles by removing the big size impurities and undissolved salt present in the solution.

This filtered solution was then used to print the channel layer over the pre-patterned Au electrodes on polyimide substrate using a Dimatix DMP 2831 ink-jet printer from Fujifilm. DuPont™ Kapton® HPP-ST with a thickness of 125 µm was chosen as the substrate because it has the advantages of good chemical and environmental resistance and can withstand a processing temperature up to 400°C without any significant change in its initial properties. The printed channel layer was dried at 100°C for 10 min on a hot plate to evaporate the solvents before annealing it at 350°C for 2 hours with the rate for heating at 5°C/min in an oven to get the Indium oxide film.

In the second system, the substrate was changed to DuPont™ Kapton® E with a thickness of 50 µm. Keeping rest of the process parameters and procedure same as described above only the annealing temperature was reduced to 300°C for 2 hours as the glass transition temperature of the DuPont™ Kapton® E is close to this temperature.

Finally, composite solid polymer electrolyte (CSPE) is printed over the structure to complete the process and was allowed to dry at room temperature to evaporate the excess solvent. This electrolyte was prepared in a similar pattern as mentioned in Dasgupta et. al. (S Dasgupta,

Gottschalk, Kruk, & Hahn, 2008) by mixing equal amounts of aqueous polyvinylalcohol (PVA, with a molecular weight of 13000-23000 g/mol, prepared at 75 °C) and aqueous potassium fluoride (KF) solution prepared separately using de-ionized water with more than 18.2 MΩ-cm electrical resistivity. The mixture was stirred for more than 12 h continuously to make it homogenous and then was filtered through 0.45 and 0.2 µm PVDF membrane filters, respectively, in order to make them printable without any nozzle coagulation. The CSPE is a crucial component of the printed transistors due to its high capacitance (1-10 $\mu F/cm^2$) and its ability to give a conformal interface even to the rough semiconductor channel film which increases the device performance due to high gating efficiency.

The Dimatix DMP-2831 ink-jet printer is used for all the printing steps. It is a small table top ink-jet printer that allows easy replacement of cartridge to facilitate printing with different kinds of inks. Each cartridge has 16 nozzles linearly spaced at 254 microns with typical drop sizes of 1 and 10 pL. Various printing parameters such as nozzle firing voltage, vacuum plate temperature, nozzle to substrate distance, drop size etc. can be adjusted to achieve the required quality of the printed drops. In our case, the printing of the Indium oxide precursor ink was done by 1 nozzle with a drop spacing of 20 µm, firing voltage of 31 V at room temperature. The nozzle to substrate distance was 0.4 mm.

3.1.2 Indium oxide nanoparticulate films

The mechanical performance of a printed device depends on several factors such as fabrication process and microstructure of the applied materials and the structure design. To understand these effects several geometrical as well as printing parameters were considered for the analysis of the mechanical behavior of printed nano-particulate (np) films targeted for thin film transistor application.

The substrate used in this study is again DuPont™ Kapton® E (50 µm thick) and the preparation and cleaning of the substrate is similar to the fabrication of the Indium oxide thin film transistors prepared by the precursor route (see section 3.1.1). The preparation of the Indium oxide nanoparticle ink (Plasmachem GmbH) is described in the following reference (Baby et al., 2015). They have successfully prepared and demonstrated a working transistor device on glass substrate by using ink-jet printed nanoparticulate Indium oxide ink as the channel material having better/comparable properties as compared to conventional transistor preparation methods.

For the preparation of Indium oxide nanoparticle ink, 1 gm of In_2O_3 nanoparticles (Plasmachem GmbH) was added to 10 ml of distilled water. 150 µl of PAANa (Sodium polyacrylate) was added into the mixture as a surfactant. For the milling of the nanoparticles 10 ml of zirconia pearls (0.2-0.3 mm diameter) were added.

A high rotating speed (8000 rpm) Dispermat (VMA GmbH) was used for 90 minutes to disperse the Indium oxide nanoparticles. During the stirring of the solution, the nanoparticle agglomerates were broken by Zirconia particles and re-agglomeration was prevented by PAANa. Subsequently, the dispersion was filtered through 5 µm, 0.45 µm, and 0.2 µm PVDF membrane filters respectively. Finally, the sintering agent NaCl (20 mM) was added to the dispersion. This ink was used for the printing of the Indium oxide nanoparticle thin films on DuPont™ Kapton® E substrate.

The chemical curing method has the advantage that the stabilizer can be removed from the nanoparticle surface which leads to formation of a conducting network without application of any heat (Golas, Louie, Lowry, Matyjaszewski, & Tilton, 2010; Grouchko, Kamyshny, Mihailescu, Anghel, & Magdassi, 2011). This makes it feasible on any kind of substrate. The nanoparticle film gets chemically sintered during the drying of the printed film as the chloride ions of NaCl detach the stabilizer molecule from the nanoparticle surface and thus allow sintering of the Indium oxide nanoparticles (Grouchko et al., 2011).

The printing of the Indium oxide np ink is done by the same Dimatix printer using 1 nozzle with a drop spacing of 20 µm and firing voltage of 31 V at room temperature. The nozzle to substrate distance was 0.4 mm. The size of the printed structure in this study is about 1x1mm except in the case of transistor geometry and orientation where the printed droplet is in the range of 100 µm X 100 µm which is comparable to the size of channel material in the printed transistor devices used in this study. The printed structures are dried at room temperature before printing of the electrolyte layer on top of it.

A schematic of samples with Indium oxide np film printed on PI is shown in figure 3.2. We have considered five different types of samples for this study. The printed samples were roughly 100 X 300 μm^2 in size.

In *type 1*, the effect of film thickness of the In_2O_3 np film printed on PI substrate on the mechanical performance of the film is studied. In_2O_3 film of multiple printed layers (3 layers, 5 layers, 7 layers and 9 layers) corresponding to different film thicknesses were ink-jet printed on PI, dried and analysed. One layer of In_2O_3 film represents one pass of printing process using a nozzle of diameter 50 µm at a jetting voltage of 40 V. The thickness of 3 layers of printed In_2O_3 film is about 300 nm measured by using a profilometer (Dektak 6M profilometer).

In *type 2*, the effect of an encapsulation of the In_2O_3 film printed on PI substrate on the mechanical performance of the film is studied. 3 layers of In_2O_3 np were ink-jet printed on PI, dried and analysed. The encapsulation was done by using the same composite solid polymer electrolyte (CSPE), which is used as the dielectric in the fabrication of the In_2O_3 thin film transistors by the precursor route in section 3.1.1.

In *type 3*, the effect of the interplay and stacking of different functional materials on the mechanical performance of the In_2O_3 nanoparticulate film printed on PI is studied. To do so, the mechanical performance of printed In_2O_3 on bare PI and on PI coated with a layer of Cr and Au is compared and analysed with and without electrolyte.

The effect of electrolyte encapsulation in addition to the effect of multiple stacking of material is also considered. The thickness of the Cr and Au is about 10 nm and 40 nm, respectively, and the films are prepared by thermal evaporation.

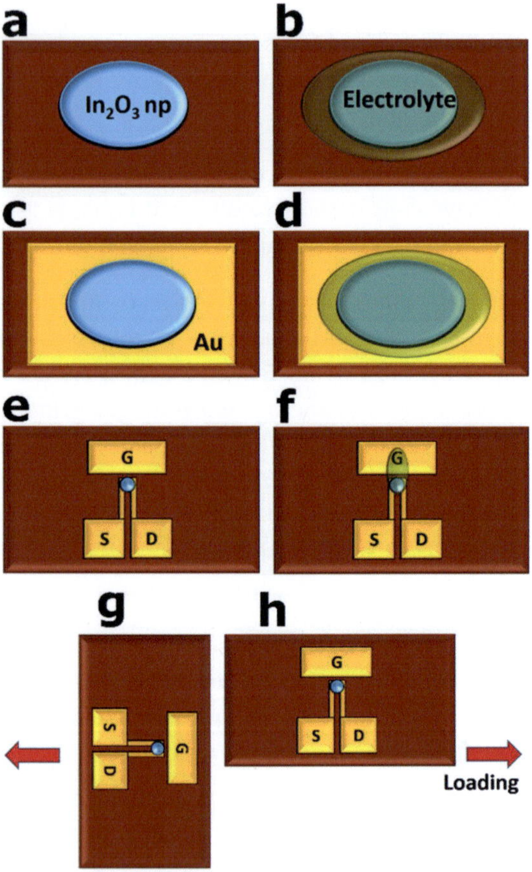

Figure 3.2: Schematic of an ink-jet printed In2O3 nanoparticulate (np) film a) on bare PI, b) film with electrolyte encapsulation c) film with Au interlayer, d) film with Au interlayer and electrolyte encapsulation e) film on transistor patterned Au electrodes, f) film on transistor patterned Au electrodes with electrolyte encapsulation, g)-h) printed transistor oriented perpendicular and parallel to the loading direction, respectively.

In *type 4*, the effect of the transistor geometry on the mechanical performance of the In_2O_3 film printed on PI has been studied. Cr and Au having thickness 10 nm and 40 nm respectively were patterned on the surface of the PI substrate by thermal evaporation using the metal mask similar to the transistor's electrode fabrication process described in section 3.1.1. The printed In_2O_3 films in the form of small droplet of size 100 µm x 100 µm were printed on the patterned Au electrodes. An electrolyte layer was printed over the printed In_2O_3 droplets to encapsulate the films completely.

In *type 5*, the effect of the transistor orientation with respect to direction of applied strain on the mechanical performance of the In_2O_3 film printed on PI is studied. Two different orientations of the transistor sample have been considered: In parallel orientation, the flow of electrons in the transistor structure was along the direction of applied strain and in perpendicular orientation the direction of flow of electrons in the transistor structure was perpendicular to the direction of applied strain.

3.1.3 BST-PMMA Capacitors

A PET substrate (Melinex® ST506; DuPont) with a thickness of 125 µm was chosen as the substrate for this study because the film is optically clear, heat stabilized to improve dimensional stability during the printing process and pre-treated on both surfaces to give enhanced adhesion of the printed material to the substrate ("Melinex ® polyester film," 2012). Figure 3.3 shows the detailed geometry of the fabricated capacitors. Preparation of Barium strontium titanate (BST)/PMMA composite ink and fabrication of all printed MIM capacitors on PET was done in the group of Dr. Joachim Binder by Morten Mikolojek, IAM-KWT, KIT.

Barium strontium titanate (BST) powder with a stoichiometric composition of $Ba_{0.6}Sr_{0.4}TiO_3$ was synthesized in a modified sol-gel process. Barium acetate (0.422mol) and strontium acetate hemihydrate (0.281 mol) were dissolved in acetic acid (30.0 mol) under nitrogen atmosphere, before titanium isopropoxide (0.703 mol) was added. The sol was diluted with water (181.8 mol), spray dried (MM-HT-ex laboratory spray dryer; Niro, Søborg, Denmark) and subsequently the metalorganic precursor was calcined at 1100 °C for 1 h. The calcined BST powder was milled and dispersed in butyldiglycol (BDG) using a laboratory stirred media mill (MiniCer; NETZSCH, Selb, Germany). The solid content of the dispersion was chosen as 39.4 wt% BST with an addition of 1.57 wt% DOLACOL D1001 (DOL; Zschimmer & Schwarz, Lahnstein, Germany) as a dispersant. The BET surface area after milling was 24.7 m^2/g (Gemini VII 2390, Micromeritics, Norcross, U.S.). A 20 vol% PMMA solution in butanone was prepared in parallel ($M_w = 1.5 \times 10^4\ g \cdot mol^{-1}$).

The composite ink was prepared by mixing the prepared 10 vol% BST dispersion and the 20 vol% PMMA solution. The following composition was obtained with a 50:50 volume ratio of BST and PMMA: 6.7 vol% BST, 6.7 vol% PMMA, 1.4 vol% dispersant, 58.6 vol% BDG and 26.6 vol% butanone (Mikolajek et al., 2015).

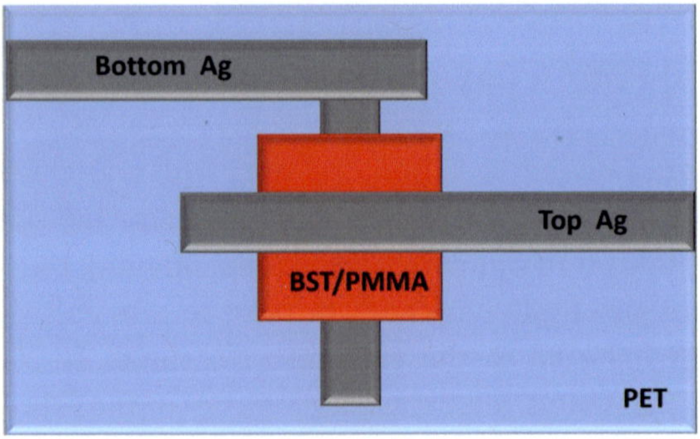

Figure 3.3: Schematic layout used for fully ink-jet printed MIM capacitors on PET substrate.

The electrodes are printed using a commercial silver ink (Silverjet DGP 40LT-15C) developed for the use on polymer substrates. The printing is done by a single nozzle piezoelectric Drop-on-Demand inkjet printer (Autodrop Professional; Microdrop, Norderstedt, Germany). Top and bottom electrodes as well as composite layer are printed with a 100 µm nozzle print head at an ejection frequency of 500 Hz. Drying temperature is set to 80 °C and 60 °C for the silver ink and the composite ink, respectively. While a single layer bottom electrode showed adequate conductivity, three layers of Ag were needed for the printing of top electrode. The BST/PMMA composite film in the ratio 1:1 was printed as a single layer to improve the homogeneity of the dielectric film (Mikolajek et al., 2015). Figure 3.3 shows the layout of the fully printed capacitors. The width of the top and the bottom electrode are 2 mm and the printed capacitors had an effective area of 2 x 2 mm^2.

The printed samples were annealed after printing of all layers was completed. To investigate the influence of the annealing temperature on the mechanical performance and the initial resistivity of the printed Ag electrodes, the printed and dried samples were annealed at 100 °C, 120 °C and 140 °C for 1 h and subsequently tested. It was found that the samples annealed at 120 °C resulted in the best properties and therefore it was chosen for all sample preparation.

3.2 Microstructural characterization

For the structural characterization of the printed Indium oxide film through precursor route, Grading Incidence X-ray diffraction was performed using a Bruker D8 Discover X-ray diffractometer with Mo-Kα radiation (50 kV, 50 mA) with a fixed incident angle of 0.4°.

A dual beam Scanning Electron Microscopy and Focused Ion Beam system (Nova Nanolab 200, FEI) is used for the microstructural evaluation of the printed layers and to monitor the surface quality, density, and thickness of the printed films. Scanning electron microscopy combined with focused ion beam milling is one of the most important techniques used in recent years as a characterization and sample fabrication tool at micro and nano scale. It consists of a SEM column along with a Ga^+ ion column which is mounted at an angle of 52° with respect to electron column. The electron column is equipped with a thermal field emission gun, which has several acceleration voltages, ranging from 1 kV to 20 kV. Usually an accelerating voltage of 5 kV and a beam current of 1.6 nA was used for microstructural examination in this study. The electron beam is used primarily for the imaging of the samples whereas the focused ion beam is used to prepare cross-sections of the samples.

The preparation of the cross-sections is done in two steps. The first step is a rough rectangular cut using a current of 5 nA, a voltage of 30 kV and a dwell time of 1 μsec. The second step is done to refine the previous cut in order to get a better finish at the surface by using same voltage and current at a longer dwell time of 3msec. The final image of the cross-section is taken using SEM at a current and voltage setting of 0.4 nA and 5 kV, respectively.

For the nanocomposite MIM capacitors the crack spacing in both the planar direction as well as on the cross-sectional direction is determined by line analysis of 10 crack lines in the SEM images of the capacitor. The surface profile and thickness of the printed droplets of the In_2O_3 np ink is analyzed using stylus profilometer (Dektak 6M profilometer, Veeco). From the measurement of the line profile, the average thickness of the printed droplet is calculated.

3.3 Electro-mechanical characterization

In-situ electrical measurement of thin film devices during the application of load is useful especially for characterization of devices fabricated on flexible substrate as it is necessary to have a consistent electrical and mechanical performance under different type of loading conditions such as stretching, bending or cyclic loading. Proper clamping of the sample and measurement of electrical data while straining the sample is challenging due to associated difficulty while sample handling (small sample size), to avoid unwanted straining of the sample while clamping and to establish good electrical contact to measure very small signal as well as collect consistent data points with high signal to noise ratio from the moving sample during straining. To study the influence of strain on the performance of the printed films and devices, different types of mechanical testing were performed on the fabricated samples and the corresponding change in the functional property of the structures is monitored.

3.3.1 Uniaxial tensile testing

The photograph of the tensile test setup manufactured by Kammrath and Weiss, Germany equipped with a 200 N load cell is shown in figure 3.4. The overall sample size for the tensile test is 4 mm x 25 mm. The ends of the PI substrate on which the transistors are fabricated were gripped on both ends and then stretched under the control of an in-build motor at a loading speed of 0.2 µm/sec up to a certain strain level. The initial length the samples between the two terminals was 8 mm.

To analyze the effect of strain on the Indium oxide transistors on PI prepared through the precursor route, the transistor properties are measured before and after the mechanical test. For the measurement of the transistor parameters an Agilent 4156C semiconductor parameter analyzer is used. A Süss MicroTec, EP6 probe station is attached to the parameter analyzer and is used for the electrical contact of the source and the drain pad of the fabricated structures. The transfer curve and the output curve of the transistor is measured though the parameter analyzer and the corresponding transistor parameters are calculated (see chapter 2.1.1).

For the in-situ electrical resistance measurement of the printed Indium oxide films on PI, special clamps are designed for tensile tests as shown in figure 3.5. Each clamp consists of two isolated copper contact pads enclosed in a plastic holder. The plastic holder insulates the two copper pads from each other. A constant current is applied over banana plugs using a Keithley Dual channel source meter (Keithley 2640 B) and the change in resistance of the films is recorded using an inhouse-developed LabVIEW program during the application of load. The mode of resistance measurement used here is 4-point resistance method as there are separate terminals for current input and voltage measurements. The advantage of using 4-point method in the resistance measurement is it reduces the effect of contact resistance in the measured resistance value for the films. For these experiments the samples were always cut to 25 mm in length and 5 mm in width.

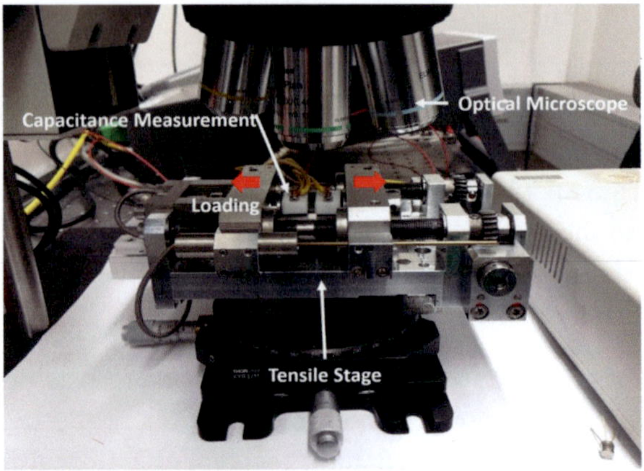

Figure 3.4: Photograph of the set-up used for tensile testing of ink-jet printed films and devices on flexible substrate.

Measurement of in-situ resistance during a tensile test is useful in detecting damage within the film due to crack formation (Lu et al., 2007). However, to estimate the effect accurately it is necessary to calculate the theoretical change in the resistance of the film due to stretching of the film without any crack formation.

The initial resistance of the film (R_0) depends on initial sample length (l_0), initial sample cross-sectional area (A_0) and the specific resistance of the material (ρ_0):

$$R_0 = \rho_0 \frac{l_0}{A_0} \tag{3.1}$$

As long as no crack formation occurs it is assumed that the specific resistance ($\rho = \rho_0$) and the material volume ($Al = A_0 l_0$) will remain constant. Considering this boundary condition, the following relation for the theoretical resistance of the film is obtained:

$$\frac{R}{R_0} = \frac{l}{l_0} = (1 + \varepsilon) \tag{3.2}$$

where $\boldsymbol{\varepsilon} = \frac{l - l_0}{l_0}$. As soon as cracking occurs within the film, the curve starts deviating from the theoretical line which is calculated from the equation (3.2).

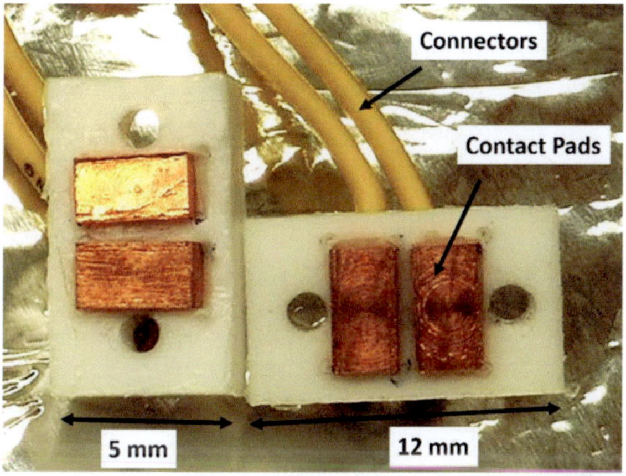

Figure 3.5: Photograph of customized clamp design used for four-point electrical characterization during tensile testing.

For the tensile testing of the ink-jet printed BST/polymer capacitors, the same tensile test setup is used. The capacitance was recorded throughout the test using a LCR meter (Agilent E4980A), connected to the tensile test equipment through customized clamps. The capacitance (Cp) and the dissipation factor (D) were measured for the printed capacitor at a frequency of 1 kHz. For the capacitors, the electrical conductivity of the bottom and top Ag electrode was measured in-situ as described above.

3.3.2 Bending fatigue testing

Bending fatigue testing is performed on printed samples to evaluate the influence of bending and cyclic loading on the performance of the In_2O_3 transistors. The setup is manufactured by CK TRADING Co. (South Korea, model name: "CK-770FET") and is shown in figure 3.6a.

The total sample size for the bending fatigue test is 4 mm x 60 mm. In this test, the sample was bent between two guiding plates and clamped on both ends. The lower plate is allowed to repeatedly move horizontally by 10 cm at a frequency of 5 Hz, keeping the upper plate fixed, and thereby inducing cyclic strain to the sample. The nature of the strain can be easily modified based on the loading conditions of the sample, whether the transistors are on the inner side (compressive) or the outer side (tensile) of the substrate. This is shown in figure 3.6 b and figure 3.6 c respectively. The maximum strain caused by bending can be calculated by Kim et. al. (B.-J. Kim et al., 2013):

$$\varepsilon = \frac{h_{Substrate} + h_{Film}}{2r} \quad (3.3)$$

where $\boldsymbol{\varepsilon}$, $h_{substrate}$, h_{film} and r are the applied strain, substrate thickness, film thickness and the bending radius, respectively. The gap between the two guiding plates is calculated as 2r and was fixed to 2.5 mm giving an applied bending strain of about 2.0 %. Due the cyclic linear movement of the lower plate three different types of deformation zones arises within the sample: two undeformed zones (Zone A), fatigue damaged zones (Zone B) and a central continuous bending zone with almost constant strain (Zone C) as shown in the figure 3.6 b and figure 3.6 c.

The two undeformed Zones A are located close to the metal grips, and, no deformation occurs in these zones as they do not bend during the test cycles. The fatigue damage Zones B are the areas which are under non-uniform strain during the bending cycles and hence show fatigue damage. The central area of the bending Zone C remains under a relatively constant and lower strain and hence does not show any fatigue damage (B.-J. Kim et al., 2013; B. Kim et al., 2014).

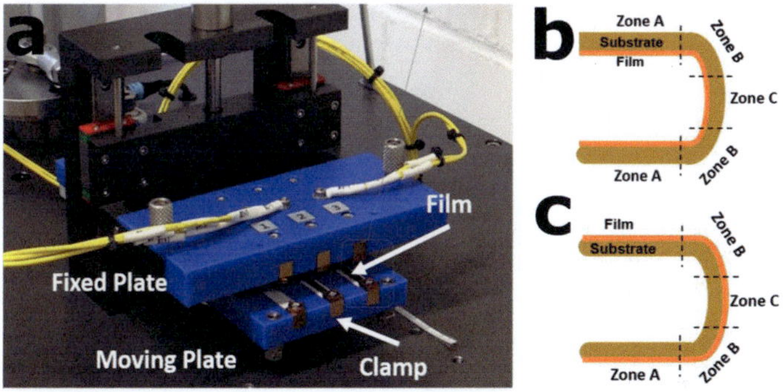

Figure 3.6: (a) Photograph of the set-up used for bending fatigue testing. The sample is mounted between the two plates. The upper plate is fixed while the lower plate can move imparting the bending fatigue stress. The schematic layout of (b) compressive and (c) tensile mode of bending fatigue testing, respectively.

Even though the applied strain in this case is not uniform over the entire area and had a gradient over the entire length, it can safely be assumed from the previous work that the transistor, if positioned in the right zone, experienced the predefined strain (Hwan Hwang et al., 2013). The transistor parameters were also recorded before and after the bending fatigue test. The bending fatigue tests were performed at a bending radius of 1.25mm (around 2 % strain) and up to 10,000 bending cycles in both tensile and compressive loading conditions.

3.3.3 Synchrotron-based tensile testing

To understand the strain transfer between the compliant substrate and the printed films, tensile tests were performed and the evolution of lattice strain within the sample was measured continuously by high flux third generation synchrotron X-Ray Diffraction (XRD) radiation at the Material Science X04 beamline facility at the Swiss Light Source, Villigen,

Switzerland (SLS MS-Powder X04A, PSI, Villigen). Synchrotron XRD radiation is chosen for the study as it has a better time resolution, brilliance and flexibility to work at variable wavelength, and as compared to lab based XRD measurements, has significantly reduced experimental time and increased resolution (J. A. Lohmiller, 2013). An Image of in-situ tensile test set-up at Material Science X04 beamline facility at Swiss Light Source, PSI, Villigen, is shown in figure 3.7.

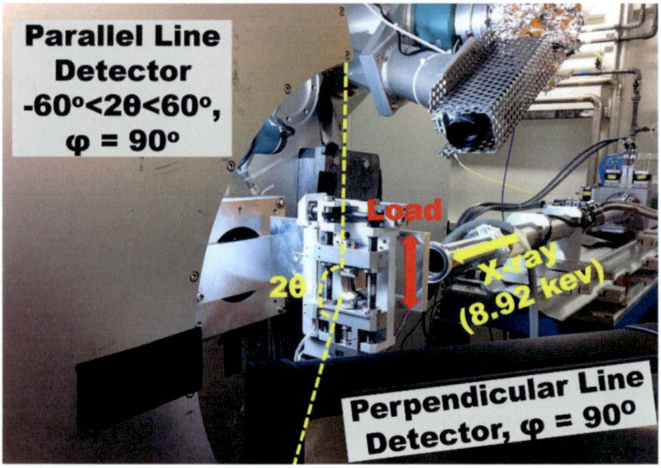

Figure 3.7: Image of in-situ tensile test set-up Material Science X04 beamline facility at Swiss Light Source, PSI, Villigen, Switzerland.

The testing procedure and data analysis were done as described by the work of Lohmiller et. al. and Olliges et. al. (J. Lohmiller et al., 2013; Sven Olliges a, Patric A. Gruber b, Vaida Auzelyte c, Yasin Ekinci c, Harun H. Solak c, 2007). The working principle of XRD is based on the following relation known as Bragg's law (Bragg & Bragg, 1913):

$$\lambda = 2 \times d_{hkl} \sin \theta_{hkl} \tag{3.4}$$

where, λ is the X-ray wavelength, d_{hkl} is the lattice spacing of specific (hkl) planes and θ_{hkl} is the corresponding Bragg's angle. When uniform strain is applied the shift in peak position is directly proportional to the change in lattice spacing and hence lattice strain in the sample can be calculated from the shift of the XRD peak position.

However, during actual straining of the samples change in shape of the peaks or reflections can be a result of several factors such as instrumental error, microstructure of the sample or due to micro strain. These factors were taken into consideration during the data analysis of the peak parameters. The X- ray beam had a cross section of around 3 x 3 mm and energy of 8.92 kev which corresponds to wavelength of about 0.71 Å. A detailed description of the test setup and data analysis is further described in Lohmiller et. al. (J. A. Lohmiller, 2013).

Two line X-ray detectors were placed vertically(Mythen microstrip, 60°<2theta<60°) (Schmitt et al., 2003) and horizontally (individual Mythen detector modules, 30°<2theta<60°) with respect to loading direction. Exemplary diffraction patterns are shown in figure3.8. The fast readout of the detectors allows continuous loading of the sample and hence avoidance of any effect due of stress relaxation in the sample. The total processing time between two X-ray diffraction patterns is about 18 sec. Peak position was determined by a Pearson VII split function (Toraya, 1990).

For the tensile test, the samples were clamped in the tensile test equipment (Kammrath and Weiss, Germany) equipped with a 200N load cell and mounted perpendicular to the direction of the monochromatic x-ray beam. The gauge length was set at 25 mm and crosshead speed was 0.8 µm/sec which results in a strain rate of 3.8×10^{-5} sec^{-1} for all experiments. The samples were pre-strained up to 0.3N to flatten the sample. The samples were elongated up to 2500 µm at the elongation speed of 0.8 µm/sec and then unloaded to 1500 µm.

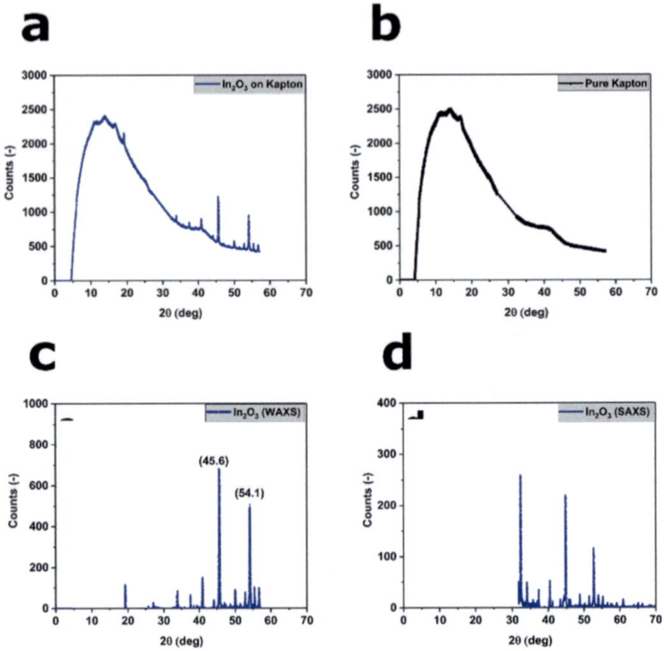

Figure 3.8: Diffraction patterns from ink-jet printed nanoparticulate In_2O_3 films on DuPont™ Kapton® E. a) As recorded diffraction pattern including background, b) background signal obtained by a blank measurement on pure DuPont™ Kapton® E film, and after background subtraction from the recorded data for c) vertical detector, and d) horizontal detector.

The total strain of the sample is measured optically by tracking markers on the sample surface, made by using a sharp gel pen within the gauge length of the sample. The instrumental effect is considered by comparing the as obtained XRD pattern of Indium oxide to the XRD pattern of a standard reference (Y_2O_3 powder). Further, to overcome the superimposed data from the polyimide substrate, a blank measurement of the DuPont™ Kapton® E substrate is recorded and subtracted from the as obtained XRD diffraction pattern. The process is shown in figure 3.8. The peak fitting for the (622) peak of In_2O_3 for both the detectors is shown in figure 3.9.

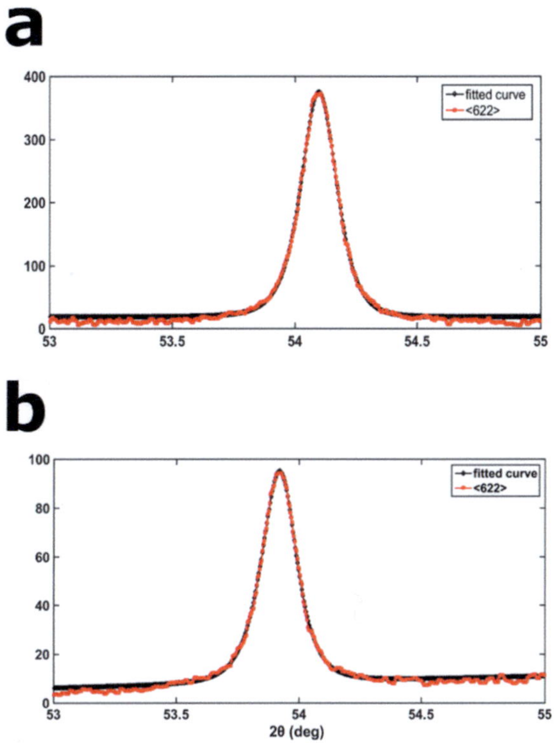

Figure 3.9: Peak fitting of the (622) In_2O_3 peak obtained from an ink-jet printed nanoparticulate In_2O_3 film on PI. a) and b) shows the PE7AM fit of the diffraction peak for both longitudinal and transverse detector, respectively.

The optical images taken during the loading of the sample were evaluated by digital image correlation (DIC) using a Matlab routine (Eberl Christoph, 2010). The optical images were taken by a Single Lens Reflex camera (Nikon D7100, 200mm Macro lens). The as obtained XRD diffraction patterns contain instrumental broadening, air scattering, and background scattering of the substrate in addition to the structural information of the material under investigation. So, the first step in the data analysis requires elimination of these effects to see the changes in the peak parameters only due to sample's structural information.

By using equation Bragg's law (equation 3.1) the lattice spacing was calculated by taking the 2θ values measured for both the detectors and the corresponding lattice strain was calculated: (Wyss, Sologubenko, Mishra, Gruber, & Spolenak, 2017).

$$\varepsilon_{hkl} = \frac{d^{n}_{hkl} - d^{0}_{hkl}}{d^{0}_{hkl}} \tag{3.5}$$

where ε_{hkl} is the lattice strain for a specific (hkl) plane, d^{n}_{hkl} is the actual lattice spacing and d^{0}_{hkl} represents the initial value of the lattice spacing before straining? The shift in lattice strain of <622> In_2O_3 peak with respect to total strain has been analysed both for the parallel and the perpendicular detector with respect to applied strain, respectively. For samples with an Au interlayer the <111> Au peak was also considered for the analysis.

4 Printed In_2O_3 TFTs

4.1 Characterization of printed transistors

To characterize the morphology of the printed In_2O_3 layer on patterned Au electrodes scanning electron microscopy is used. Figure 4.1 shows the quality of the indium oxide film printed through two different systems as has been described in the section 3.1.1. It has been found out that the quality of the In_2O_3 film fabricated on DuPont™ Kapton® E is superior to the first system which has been printed on DuPont™ Kapton® HPP-ST substrate.

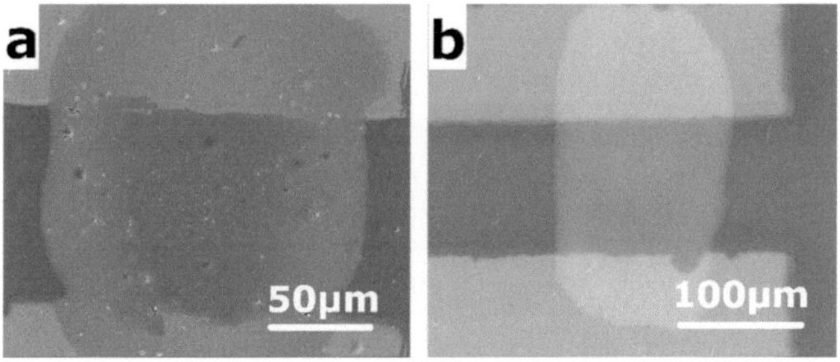

Figure 4.1: SEM images of the ink-jet printed In_2O_3 film prepared through precursor route and printed on patterned Au electrodes on a) DuPont™ Kapton® HPP-ST substrate and annealed at 350°C for 2 hr. b) DuPont™ Kapton® E and annealed at 300°C for 2 hr.

In the first system, the printed indium oxide films are dense and uniform however there are some hillocks all over the film surface. It has been found out that these hillocks deteriorate the mechanical performance of the transistors which is discussed later in this chapter. In the second system, no such artefacts are found, and the film also looks homogenous and dense. Therefore, the second system was used for the fabrication of the transistors and the detailed description of their mechanical and electrical performance is discussed in the following section. For the structural characterization of the printed indium oxide film X-ray diffraction was performed on the In_2O_3 film deposited and annealed on a glass substrate. A glass substrate was used in this case instead of PI because the huge background of the PI made it difficult to analyse indium oxide peaks with a standard laboratory X-ray diffractometer. The XRD measurements were performed using a Philips X-ray diffractometer with Cu-Kα radiation with voltage and current set to 45 kV and 40 mA, respectively. It is confirmed that Indium oxide is formed from the precursor solution after annealing the film at 300°C for 2 hr as shown in figure 4.2.

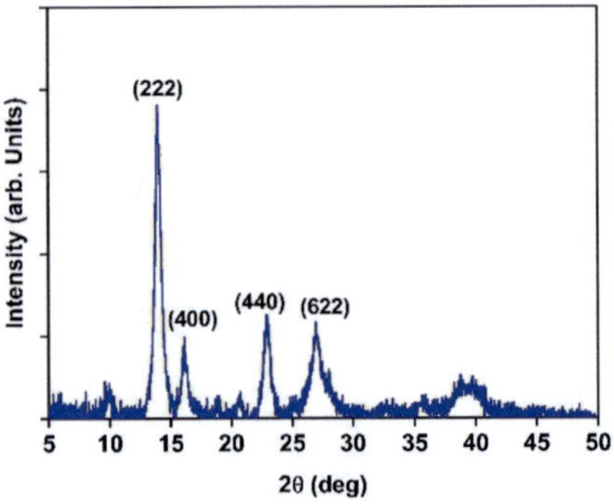

Figure 4.2: XRD pattern of the printed In_2O_3 film from precursor route after annealing at 300°C for 2 hr. XRD peaks are indexed with respect to the standard indium oxide phase ICSD No. 06-0416.

The measured XRD data is compared to the reference ICSD No. 06-0416 for In_2O_3 to index the XRD peaks Electrical characterization of the printed transistors was performed at room temperature and at normal atmospheric conditions. The transfer curve is obtained by applying a constant drain voltage of 1V and a sweeping voltage of +1V to -1V between gate and source electrode. The output curve is obtained by applying a gate voltage of 0.5 V and a sweeping voltage of 0 V to 2 V is applied between drain and source electrode. Figure 4.3 represents the typical transfer and output curve of the printed transistor.

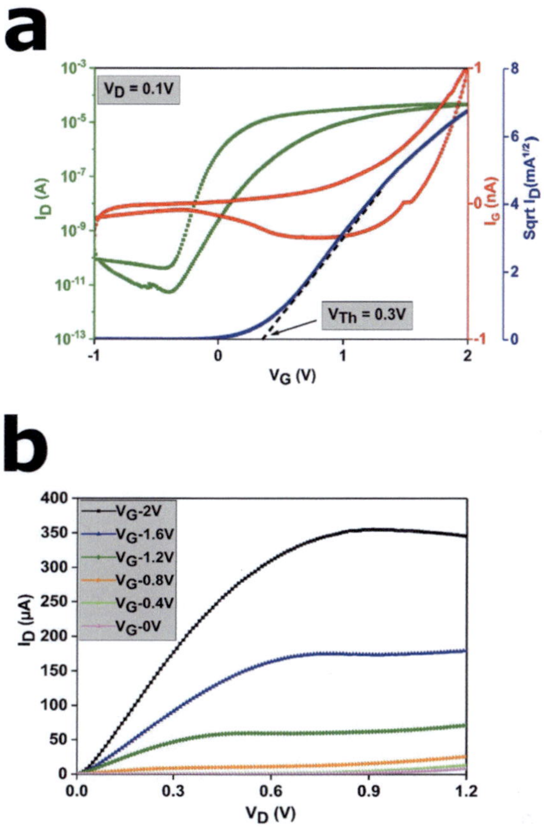

Figure 4.3: a) Transfer and b) Output curves of the ink-jet printed transistor on DuPont™ Kapton® E using indium oxide precursor and annealing at 300 °C for 2 hr. The transfer curve shows the drain current (I_D, green line), the gate current (I_G, red line), the square root of drain current ($I_D^{1/2}$, blue line) as a function of gate voltage (V_G). The output curves show the drain current with variable gate voltage from 0 to 2 V with a step size of 0.4 V as a function of drain voltage (V_D).

The transfer curve shows that when a negative gate voltage is applied the transistor device is at off state. When the gate voltage approaches the threshold value of about 0.3 V, the device current starts increasing and attains a saturation value at around 2 V. When the gate voltage is below

0.3 V the depletion layer is formed in the semiconductor layer which prevents flow of electrons form source to gate even for the application of a drain voltage. However, as soon as the gate voltage approaches the threshold value an electrical double layer is formed at the semiconductor electrolyte interface which assists the flow of electrons from source to drain. The output curve shows that source-drain current approaches a saturation value at around 0.5 V irrespective of the increase in the gate voltage. The maximum drain current is obtained at a gate voltage of 2 V.

From these plots, the following transistor parameters are obtained: maximum drain current of 327µA at an applied gate voltage (V_G) of 2 V, threshold voltage (V_T) of 0.3 V and an I_{ON}/I_{OFF} ratio in the order of 10^7. From the positive threshold voltage, we can conclude that the device is in accumulation mode. The transistor parameters are calculated by using the following equation, where $I_{D,sat}$ is the saturated drain current, W and L are the width and length of the channel layer, respectively, C is the double layer capacitance per unit area for the device which is taken as 4.63 µF/cm² from our previous work (Garlapati et al., 2015):

$$I_{D,sat} = \left(\frac{W}{2L}\right) C\mu_{sat}(V_G - V_T)^2 \qquad (4.1)$$

The field-effect mobility μ_{sat} is calculated as 70.61 cm²/Vs, and the subthreshold swing as 70.0mv/decade. All the transistor parameters are also listed in Table 4.1. The measured leakage current is around 1-2 nA. The small hysteresis in the transfer curve is due to the in-plane transistor geometry. High quality indium oxide transistors on a flexible substrate are successfully fabricated though a precursor route and with high electrical performance. This is concluded from the obtained high values of drain current at a low applied voltage of 2 V (due to the application of high capacitance electrolyte), positive threshold voltage (the device is not at ON state), low sub-threshold swing (close to its theoretical value which is 68 mv/dec) and a high field effect mobility almost comparable to indium oxide transistors reported on silicon substrates (Garlapati et al., 2013, 2015).

Table 4.1: Calculated transistor parameters for precursor based In_2O_3 transistors annealed at 300 °C.

Maximum drain current	327 µA
Field effect mobility	70.61 cm^2/Vs
Threshold Voltage	0.3 V
Sub-threshold swing	70 mv/dec
I_{On}/I_{Off} ratio	10^7

4.2 Tensile testing of printed transistors

To understand the influence of external strain on the performance of ink-jet printed indium oxide transistors on PI substrate, uniaxial tensile tests were performed. The transistor parameters were determined before and after the mechanical test.

As shown in figure 4.4 a, the hillocks present in the In_2O_3 printed on DuPont™ Kapton® HPP-ST substrate, initially do not affect the device electrical performance as we get a very high drain current of 422 µA at a small operating voltage of 2 V with a low threshold voltage of 0.5 V. However, when they are subjected to mechanical strain they failed at a very low tensile strain of less than 2 % as the presence of hillocks leads to crack initiation and further crack propagation within the oxide layer.

This can be confirmed from the SEM images of the indium oxide film in figure 4.4 b and figure 4.4 c subjected to 0 % and 2 % of tensile strain respectively. Therefore, rest of the work include studies on transistors printed on DuPont™ Kapton® E and annealed at a temperature of 300 °C for 2 hours. Moreover, two different orientations of the printed transistors are also considered for this study as has been described in the experimental section, Parallel Orientation (flow of electron parallel to the direction of applied load and Perpendicular Orientation (flow of electron perpendicular to the direction of applied load). To understand the effect of tensile strain, the output curve of the transistor is measured while sweeping the drain voltage from 0-2 V and keeping the gate voltage constant at 2 V. In figure 4.5, the change in maximum drain current at a constant applied voltage of 2 V with respect to applied strain for parallel and perpendicular orientation is shown, respectively.

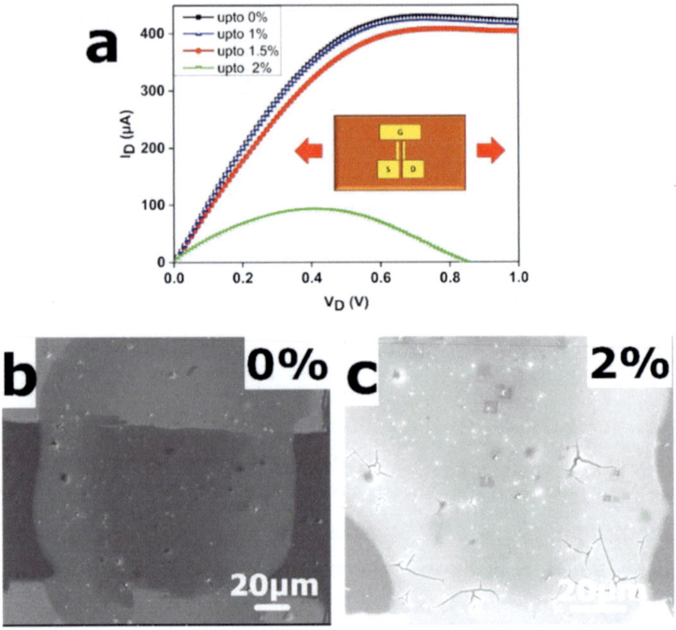

Figure 4.4: a) Output curves of the printed transistor on DuPont™ Kapton® HPP-ST substrate at a constant gate voltage of 2V and at variable applied tensile strain up to 2 %, b) and c) SEM images of the ink-jet printed In_2O_3 film on DuPont™ Kapton® HPP-ST substrate upto 0 % and 2 % tensile strain respectively.

For parallel orientation, it is observed that the maximum drain current obtained at an applied voltage of 2 V is stable up to 2 % of tensile strain. At around 2.5 % of tensile strain the maximum drain current starts decreasing with increase in applied tensile strain and at around 3 % of tensile strain the drain current value decreases to zero indicating complete failure of the indium oxide transistor.

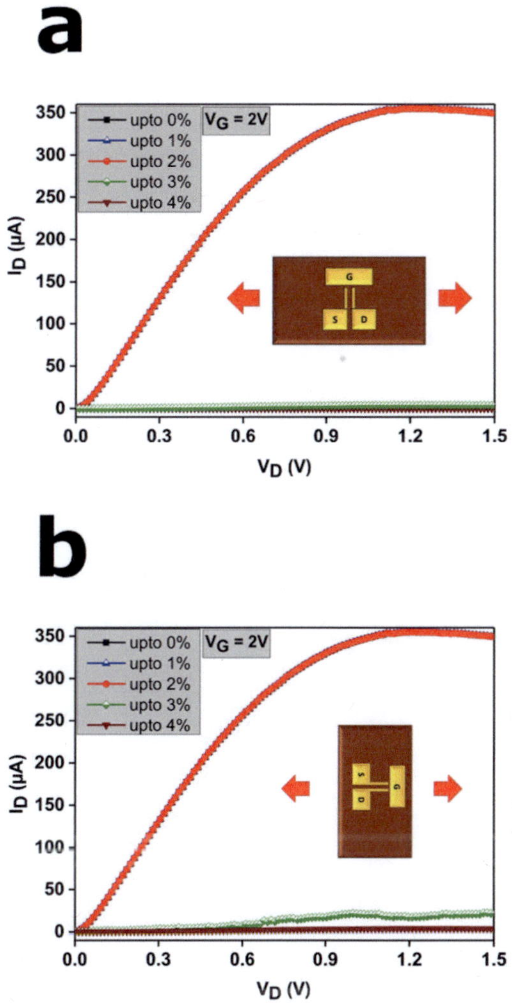

Figure 4.5: a) & b) variation of drain current (I_D) vs drain voltage (V_D) at constant gate voltage of 2 V for ink-jet printed In_2O_3 transistors at different applied strain in parallel and perpendicular loading orientation, respectively.

For perpendicular orientation, it is observed that the maximum drain current obtained is stable up to 2 % of tensile strain similar to the previous case and then starts decreasing with increase in applied strain. The change in the performance parameters of the printed indium oxide transistor with respect to externally applied strain is similar in both cases, irrespective of the orientation. This decrease in measured drain current with increase in applied strain is due to crack formation within the oxide layers as can be seen in the SEM images of the oxide layers after the application of strain as shown in figure 4.6. Initially the indium oxide film printed on passive electrodes looks uniform and dense without any crack or voids as shown in figure 4.6a. Up to 2 % tensile strain no cracks were observed as shown in figure 4.6b.

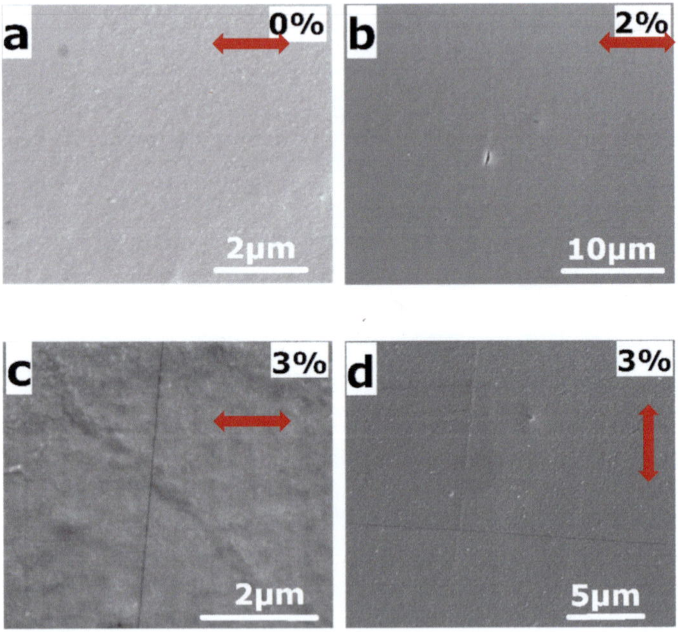

Figure 4.6: SEM images of crack formation in ink-jet printed In_2O_3 layer at different tensile strain level a) 0 %, b) 2 %, c) 3 %, for parallel oriented transistor and d) 3 %, for perpendicular oriented transistor.

For higher strain, channel cracks develop across the indium oxide film in a direction perpendicular to the direction of applied strain in both cases irrespective of the orientation of the devices. In parallel orientation, as the propagation of cracks in this case is perpendicular to the direction of the flow of electrons, it was expected that a sudden decrease in maximum drain current with increase in strain level leading to failure of the device should occur.

However, for perpendicular orientation, as the crack propagation is along the direction of the flow of electrons, it was only expected to see a gradual decrease in current with increasing crack density. To understand this phenomenon and to compare the failure mechanism of the printed transistors with the printed films, tensile tests were performed on indium oxide films directly printed on bare DuPont™ Kapton® E and annealed under similar conditions as before. The change in device microstructural and the electrical parameters are recorded and compared as shown in figure 4.7 and figure 4.8 respectively.

To understand this phenomenon and to compare the failure mechanism of the printed transistors with the printed films, tensile tests were performed on indium oxide films directly printed on bare DuPont™ Kapton® E and annealed under similar conditions as before. The microstructural and the electrical parameters changes are recorded and compared as shown in figure 4.7 and figure4.8 respectively.

From figure 4.7a, it is observed that for parallel orientated films the resistance of the indium oxide films increases sharply around 3 % of tensile strain. This is due to formation of through-thickness cracks in the film surface as shown in figure 4.8d and 4.8e. For parallel orientation, the failure mechanism of the indium oxide films printed is similar. From the figure 4.7b, it is observed that for perpendicular oriented films the increase in the resistivity of the Indium oxide films apart from the theoretical resistivity increase, starts around 3 % of tensile strain and then continue to increase only gradually till 20 % of tensile strain. However, the decrease in the film resistance is gradual because the cracks were parallel to the flow of electrons and thereby only leading to a minor increase in the film resistance with increasing crack density, finally leading to a film failure around 20 % of tensile strain as shown in figure 4.8a, figure 4.8b and figure 4.8c.

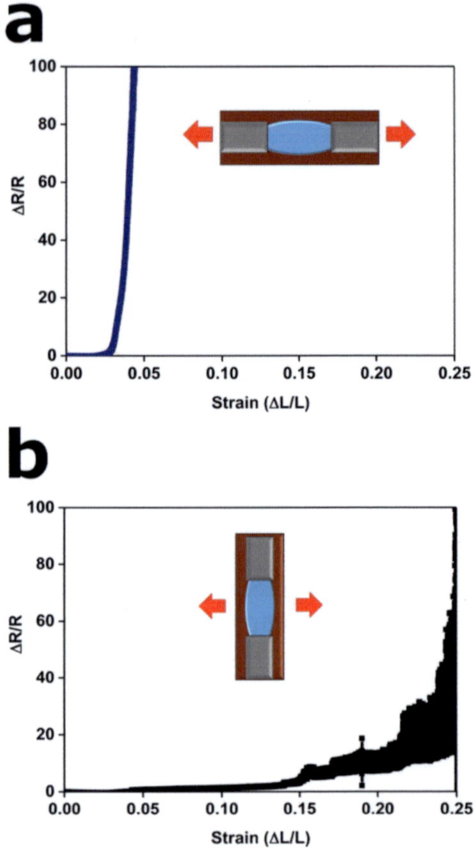

Figure 4.7: In-situ resistivity change on the ink-jet printed In_2O_3 film on DuPont™ Kapton® E. The normalised resistivity increases with increasing tensile strain, a) parallel and b) perpendicular orientation of the film, respectively.

Based on these results it can be concluded, that the electrical performance of the indium oxide films indeed depends on the orientation of the film with respect to electron flow and loading direction and further, that the crack formation in the indium oxide films is not the only cause for the sudden failure of the printed transistors. Possible further failure mechanisms will be discussed later in this chapter.

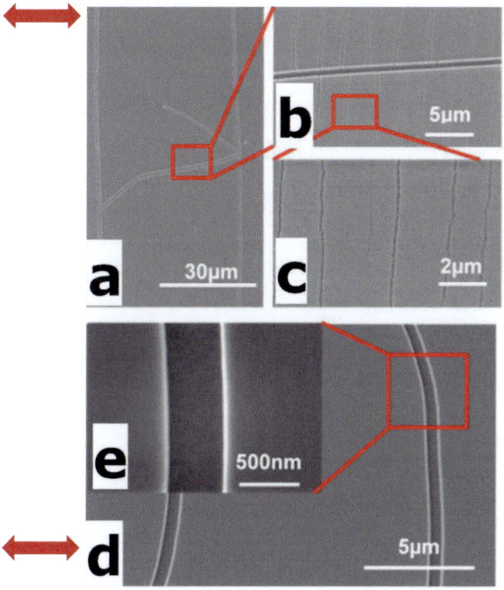

Figure 4.8: Crack morphology within the ink-jet printed In_2O_3 layer oriented a), b), c) in perpendicular orientation and d), e) in parallel Orientation with respect to applied tensile strain as shown by red arrow mark.

4.3 Bending fatigue testing on transistors

Bending fatigue tests were conducted to examine the variation in transistor parameters under cyclic tensile or compressive loading. The maximum applied strain for the bending fatigue tests was limited by the minimum bending radius achievable for the bending fatigue test equipment (1.25 mm) and the thickness of the polyimide substrate (50 µm). Therefore, the maximum strain that can be applied to the printed transistors is 2 %, which is below the failure onset strain for the printed transistors.

The fatigue tests were also conducted for both, the parallel and perpendicular orientated transistors. Output curves of the printed transistor for parallel and perpendicular orientation under cyclic tensile and cyclic compressive loading are shown in figure 4.9. The indium oxide transistors were fully operational up to 10,000 bending cycles when subjected to cyclic tensile and compressive loading with a maximum strain of 2 %.

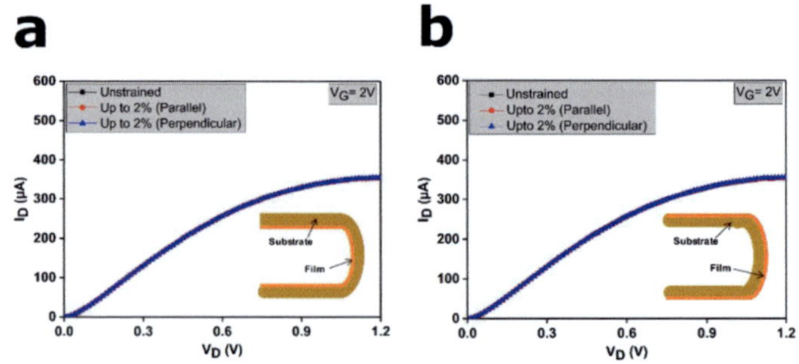

Figure 4.9: Output curves of the ink-jet printed transistors at constant gate voltage of 2V for parallel and perpendicular orientation up to a) 2 % cyclic tensile bending and b) 2 % cyclic compressive bending strain, respectively.

4.4 Discussion

In general, mechanical properties of thin films are different than for bulk materials due to their microstructural and geometrical constraints and it is not possible to simply apply classical mechanical testing methods to extract these properties (O. Kraft & Volkert, 2001; Nix, 1989; Olliges et al., 2008; Spaepen & Shull, 1996). Further, use of ink-jet printing technique for the film deposition process results in a nanoporous microstructure. Ink-jet printed films has been reported to have lower fracture strain as compared to evaporated or sputtered fully dense films when subjected to uniaxial tensile testing because of the presence of nanopores in the film, which act as stress concentrators and aggravate the formation of cracks (B.-J. Kim et al., 2013; B. Kim et al., 2014; J.-H. Lee, Kim, Kim, & Joo, 2012; Sim, Won, & Lee, 2012). On the other hand, it has also been reported that the increase in electrical resistance of the ink-jet printed films is lower compared to evaporated films when subjected to cyclic loading (H.-Y. Lee et al.,2010).

The printed Indium oxide transistors are consisting of a brittle Indium oxide layer, a ductile Au layer, and a polymer electrolyte layer. Therefore, there are several critical factors that come under consideration for understanding the mechanical behavior of these printed transistors on flexible substrates. One of them are sharp edges in the transistor geometry which act as stress concentrator due to edge effect when subjected to external strain (Sottos, Kandula, Geubelle, & Mechanics, 2005; H. Yu, 2001). The second factor is stacking of multiple materials having variable film thickness, different coefficient of thermal expansion and degree of interfacial adhesion. The third factor is the effect of encapsulation of the oxide layer with the electrolyte which result in ratcheting effects in the transistor structure (Huang, Suo, & Ma, 2002). All these parameters can change the mechanical properties of the device significantly.

The development of electro-mechanical testing is necessary for the systematic investigation of the effect of uniaxial and cyclic tensile and com pressive loading on the reliability of the printed indium oxide transistors on flexible substrate. From the observations so far, it can be concluded that the quality of the printed indium oxide films plays an important role in the mechanical performance and reliability of thin film devices. The presence of hillocks or artefacts in the printed oxide layer do not aggravate changes the initial electrical device performance, however, device failure occurred at a low tensile strain of 2 %. It has been shown that the failure strain of the device can be improved further by developing oxide films devoid of such hillocks and artefacts by using better inks and better substrate surface qualities.

Further, based on the experiments with different orientation of the transistors with respect to applied strain, the difference in the mechanical behavior of printed thin films and thin film transistors confirms that in case of printed transistors, the formation of cracks in the oxide layer is not the only factor responsible for the failure of the devices. The possible explanation for the sudden decrease in drain current irrespective of the orientation of the transistors is most probably due to delamination of the oxide layer right above the electrodes due to the localized strain concentration at the electrode edges. This leads to reduction or loss of the electrode-channel contact and hence decreases in the output current of the device (H. Yu, 2001). This hypothesis has not been confirmed experimentally by SEM as cross section cannot be examined at tension.

In this study, the effect of mechanical strain on the electrical performance of printed indium oxide transistors on flexible substrate has been investigated. It has been found out that transistors fabricated by a precursor route are having a very good electrical performance with a field-effect mobility of 70.61 cm^2/Vs and the I_{ON}/I_{OFF} ratio is in the order of 10^7 at a very low applied voltage and has a reasonably good mechanical flexibility. The reported transistors were fully operational up to 10,000 cycles at an applied bending strain of 2 % under both tensile and compressive loading conditions. Under uniaxial tension the transistors were fully operational up to 2 % strain and then start deteriorating due to formation of cracks within the oxide channel layer and due to delamination at the interface between the Au electrode and the indium oxide layer. It is evident that the morphology of the indium oxide film and the surface quality of the substrate are determining the strainability of the flexible transistors. A more homogenous and inclusion free channel layer, shorter channel length and top gate geometry will help in enhancing the mechanical performance and reliability of these ink- jet printed flexible indium oxide transistors.

5 Printed nanoparticular In_2O_3 films

In this section, the results of in-situ tensile testing of ink-jet printed indium oxide nanoparticulate (np) films on DuPont™ Kapton® E polyimide substrate (PI) are presented. The influence of applied strain on the lattice strain evolution of the np films is investigated using synchrotron X-ray diffraction.For this study several printing and geometric parameters have been considered for analysis such as film thickness, effect of encapsulation, effect of multiple layering of materials, effect of transistor geometry and its orientation with respect to an applied strain.

Figure 5.1 shows the SEM images of the printed and dried indium oxide nanoparticulate films on DuPont™ Kapton® E at a lower and higher magnification, respectively. The film quality looks porous with particle to particle connectivity, which is essential for high electrical conductivity of the Indium oxide film.

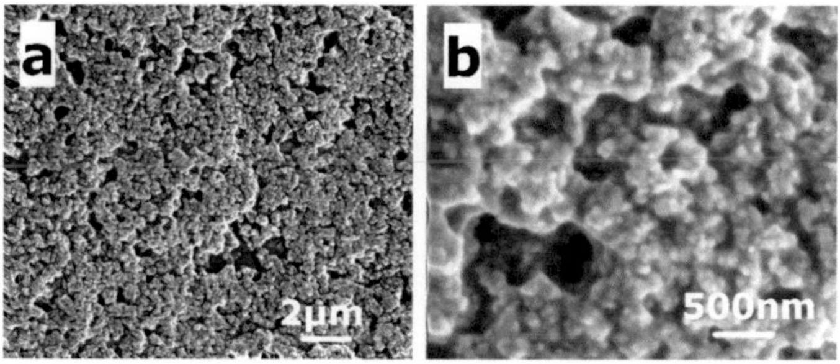

Figure 5.1: SEM images of a printed indium oxide nanoparticulate film on DuPont™ Kapton® E at a) lower and b) higher magnification.

In_2O_3 np films were ink-jet printed on polyimide, dried and in-situ tensile tests were performed at the Materials Science beamline of the Swiss

Light Source (Villigen, Switzerland). The shift of the <622> peak ($2\theta \sim 54.3^{o}$) with respect to an applied tensile strain was measured and the corresponding lattice strain was calculated for each sample using the method described in section 3.3.3. Using two wide angle line detectors, the lateral and perpendicular lattice strain was obtained.

5.1 Effect of film thickness

In this section, the effects of the thickness of the printed In_2O_3 films are studied. Results of the shift in <622> lattice strain of the material with different thickness with respect to true strain are shown in figure 5.2. Figure 5.2a and 5.2b represent the shift in lattice strain for the longitudinal (Parallel) and the transverse (longitudinal) detector, respectively. The black, red, green and blue lines represent the In_2O_3 droplet printed on bare polyimide with thickness 9L, 7L, 5L and 3L, respectively, where L correspond to number of printing passes.

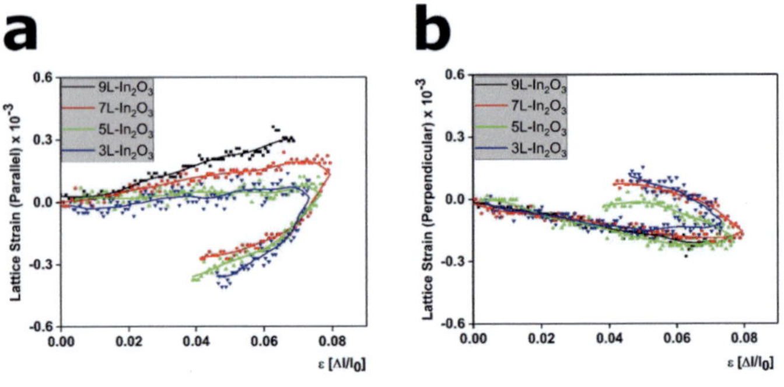

Figure 5.2: a) Longitudinal and b) transverse lattice strain evolution of the ink-jet printed In_2O_3 film on DuPont™ Kapton® E as a function of true strain with variable film thickness.

In figure 5.2a, a few important trends are observed. First, during the loading of the printed In_2O_3 films, the lattice strain in the In_2O_3 films for different thicknesses is increasing with increase in loading and during the unloading it starts decreasing irrespective of the thickness of the material. Secondly, the increase in lattice strain with the elongation of the sample is higher for thicker In_2O_3 films (7L and 9L) as compared to thinner In_2O_3 films (3L and 5L). The maximum lattice strain ε_{max} achieved by the thicker films is around 3 x 10^{-4} whereas the thinner sample shows barely any strain transfer (<1 x 10^{-4}).

Third, irrespective of the thickness of the material, during the unloading of the sample, the lattice strain of the sample is decreasing instantaneously in a similar manner and rapidly becomes compressive. Since it's a biaxial strain measurement we observe the same trends in the transverse detector (perpendicular direction) just with opposite sign due to Poisson contraction, irrespective of the thickness of the material. The signal quality of transverse detector is usually inferior to that of longitudinal detector, which is technically given by the different type of detector modules.

5.2 Effect of film encapsulation

Here, the effect of electrolyte encapsulation of the ink-jet printed In_2O_3 np film is studied. Results of the shift in <622> lattice strain of the In_2O_3 np films with respect to true strain are shown in figure 5.3.

The thickness of the Indium oxide films is always 3L. In figure 5.3a, the lattice strain in the In_2O_3 films is again increasing with increase in elongation of the In_2O_3 films/substrate and then starts decreasing with the unloading. The blue line represents shift in <622> lattice strain of In_2O_3 np films without electrolyte encapsulation and the green line represents the same film with encapsulation. The maximum lattice strain during loading is similar for both samples, however, during unloading, the instantaneous accelerated reversion of the lattice strain observed for the bare indium oxide nanoparticulate film is significantly reduced for the encapsulated film. This effect of electrolyte encapsulation is also observed in the transverse direction, again with opposite sign.

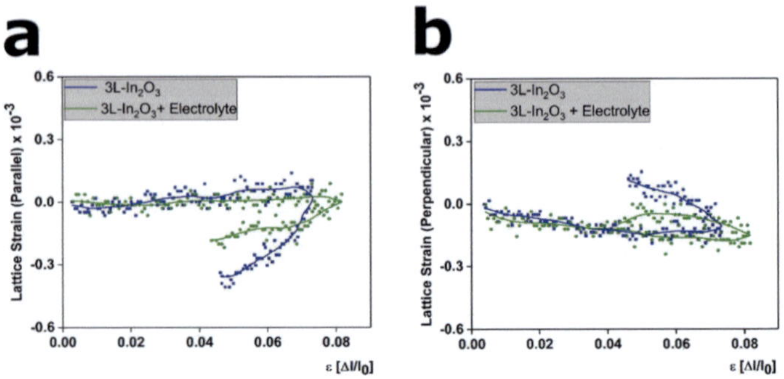

Figure 5.3: a) Longitudinal and b) transverse lattice strain evolution of the ink-jet printed In_2O_3 3 layer film on DuPont™ Kapton® E as a function of true strain with and without electrolyte encapsulation of the film.

5.3 Effect of multilayer

In this section, the effect of the layering of different materials on the mechanical properties of the In_2O_3 film printed on PI is studied. In this case three different samples were tested and the shift in lattice strain is compared to see the effect of multiple layering of different materials. Results of the shifts in <622> lattice strain of the In_2O_3 film and <111> lattice strain of the Au film with respect to the applied true strain are shown in figure 5.4. The blue lines represent the In_2O_3 film printed on bare PI, the yellow line represents the In_2O_3 film printed on PI having an interlayer of Cr/Au and the green line represents the In_2O_3 film printed on Cr/Au and with an encapsulation of polymer electrolyte on top of it.

In figure 5.4a, the <622> lattice strain of In_2O_3 np film is increasing with loading and then it starts decreasing with the unloading of the sample as has been observed previously. Furthermore, the lattice strain evolution in the Indium oxide film in all the three samples is similar during the loading part of the curve and the maximum lattice strain value achieved by the In_2O_3 is insensitive to the addition of the Cr/Au interlayer.

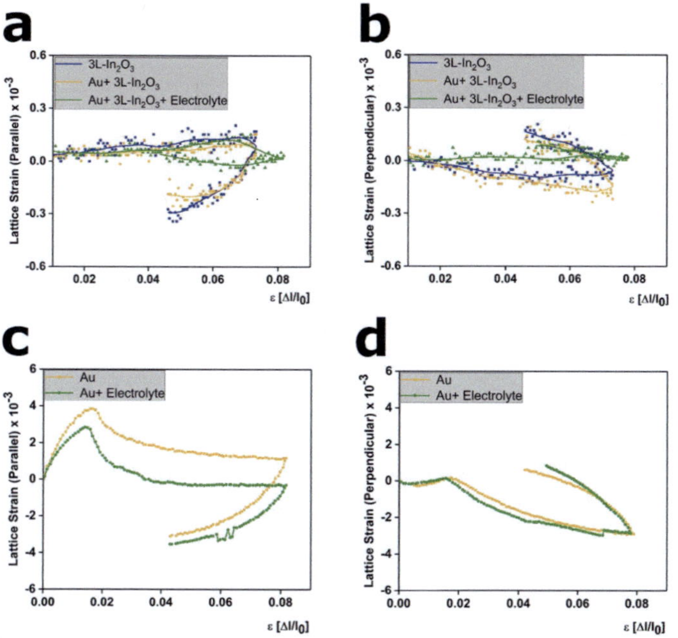

Figure 5.4: a), c) Longitudinal and b), d) transverse lattice strain evolution of the printed In_2O_3 and Au films respectively as a function of true strain with and without electrolyte encapsulation of the films.

However, during unloading the effect of encapsulation is again visible. The lattice strain evolution of the In_2O_3 film in the transverse direction i.e. in figure 5.4b is compressive while loading and then it starts relaxing during the unloading section. The effect of electrolyte encapsulation is also observed here. For the Au film (figure 5.4c), the <111> lattice strain evolution in both the samples i.e. Au film of PI and Au film on PI with electrolyte encapsulation increases during the loading of the sample up to 2 % of tensile strain and then starts decreasing till it reaches a plateau. During unloading, the lattice strain again decreases and becomes compressive until it reaches a second plateau in compression. Please note that the described procedure only evaluates relative lattice strains with respect to the initial lattice spacing, as the absolute lattice spacing cannot be obtained by the transmission diffraction geometry. The presented lattice strain evolutions thereby neglect residual strains in the films.

The level of the strain plateaus would have to be shifted by the residual strain value, as for evaporated and dense Au films significant tensile residual strains are expected. An estimate for the unstrained lattice parameter could be obtained by evaluating the center line between the strain plateaus in tension and compression assuming symmetric mechanical behavior in tension and compression (Wyss et al., 2017). This procedure would shift the lattice strain curve to higher values as expected for tensile residual strains in the Au film. Overall, the range in lattice strain values achieved in the Au films is at least one order of magnitude higher as compared to the np indium oxide films. This reflects a much higher strain transfer between the substrate and the dense and well bonded Au film. Electrolyte encapsulation does not have a significant effect is the strain evolution in the Au film.

The lattice strain evolution in the Au film in the transverse direction is constant until the strain drop in the longitudinal lattice strain is reached and subsequently becomes compressive. During unloading the lattice strain in the transverse direction relaxes to its initial value. An effect of the electrolyte encapsulation is also not noticeable here

5.4 Effect of transistor geometry

In this section, the effect of the transistor geometry on the mechanical properties of an ink-jet printed In_2O_3 np droplet having a similar size as for the transistor design (100 µm x 100 µm) on PI is studied. Again, three different samples, bare Indium oxide droplet, Indium oxide droplet printed on patterns Cr/Au electrodes and Indium oxide droplet on Cr/Au electrodes with polymer electrolyte encapsulation were tested. Results of the shift in <622> lattice strain of the In_2O_3 np droplet and results of the shift in <111> lattice strain of the patterned Au electrodes with respect to true strain are shown in figure 5.5. The blue line represents the In_2O_3 droplet, the yellow line represents the bare Au electrodes, the brown line represents In_2O_3 droplet printed on patterned Au electrodes and the green line represent the In_2O_3 np droplet printed on Au electrodes and with an encapsulation of polymer electrolyte on top of it.

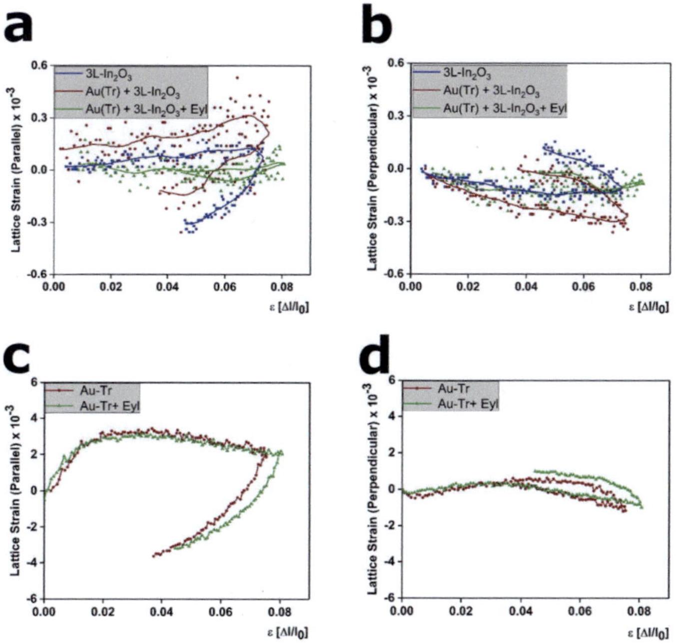

Figure 5.5: a), c) Longitudinal and b), d) transverse lattice strain evolution of the printed In_2O_3 and Au structures as a function of true strain respectively.

As can be seen in figure 5.5a and figure 5.5b, the signal-to noise ratio for the In_2O_3 np droplet samples becomes inferior due to the significant decrease in the scattering volume. The strain evolution in each individual sample is similar to previous cases discussed so far. However, the lattice strain maximum in the In_2O_3 np droplets is higher as compared to In_2O_3 np films. The effect of encapsulation is still observed. In figure 5.5c and figure 5.5d, the lattice strain evolution for the <111>Au peak of the Au electrodes shows only a minor and subtle decrease in the lattice strain evolution and no strain drop is observed.

5.5 Effect of transistor orientation

In this section, the effect of orientation of the transistor with respect to the direction of loading is studied. Two different samples were tested and the shift in lattice strain has been compared. Results of the shift in <622> lattice strain of the In_2O_3 np droplet and of the shift in <111> lattice strain of the patterned Au electrodes with respect to true strain are shown in figure 5.6.

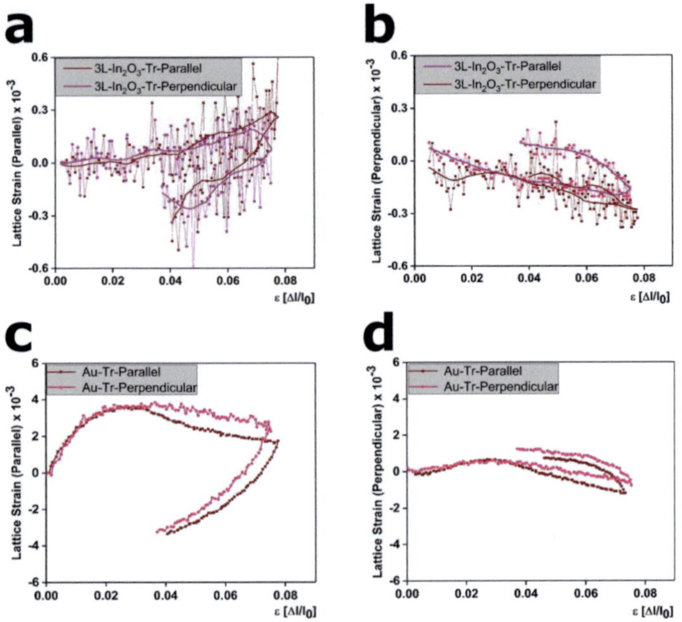

Figure 5.6: a), c) Longitudinal and b), d) transverse lattice strain evolution of the printed In_2O_3 and Au respectively on DuPont™ Kapton® E as a function of true strain because of transistor orientation with respect to loading direction.

The brown line represent the In_2O_3 droplet printed on patterned Au electrode orientated parallel to the direction of loading and the magenta line represents the transistor sample orientated perpendicular to the direction of loading. No significant difference can be observed between the lattice strain evolutions of the nanoparticulate Indium oxide droplet and the Au electrodes for the parallel and perpendicular orientation.

5.6 Discussion

The lattice strain evolution in the In_2O_3 np film and droplet can be explained by a model which is also schematically shown in figure 5.7. For thinner films, In_2O_3 nanoparticles behave as particles that are loosely connected to each other. Therefore, they behave like particles of sand but with connections. During the elongation of the sample, these particles reorient themselves with respect to each other and the lattice does not accommodate any strain from the substrate, while retaining the particle-to-particle connection. However, the reorientation of the nanoparticles is constricted when the thickness of the film increases due to the increasing number of particles and lower number of degrees of freedom for particle movement.

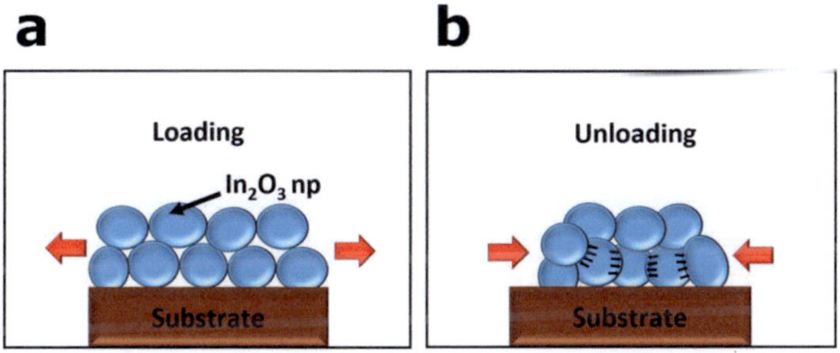

Figure 5.7: Schematic model to explain the lattice strain evolution of ink-jet printed In_2O_3 film on a polymer substrate during a) loading and b) unloading of the sample.

Therefore, for thicker films, the lattice spacing of the In_2O_3 particles starts accommodating strain from the substrate due to increasing frictional forces and increasing number of static connections between the nanoparticles. However, during unloading of the sample the nanoparticle connections are under compression which results in higher friction forces and transfer of forces across the interface of the connected particles and hence the particles instantaneously have to accommodate strain.

In the study of the effect of electrolyte encapsulation on the lattice strain evolution of the In_2O_3 np film, it has been observed that the strain behavior in In_2O_3 film with and without electrolyte encapsulation is similar during the loading of the sample i.e. under tension. However, during unloading, a significant effect of the encapsulation is observed, and the compressive strain accommodated by the In_2O_3 film is reduced in the case of encapsulation. One of the possible explanations for the difference in the strain transfer mechanism is shown in figure 5.8.

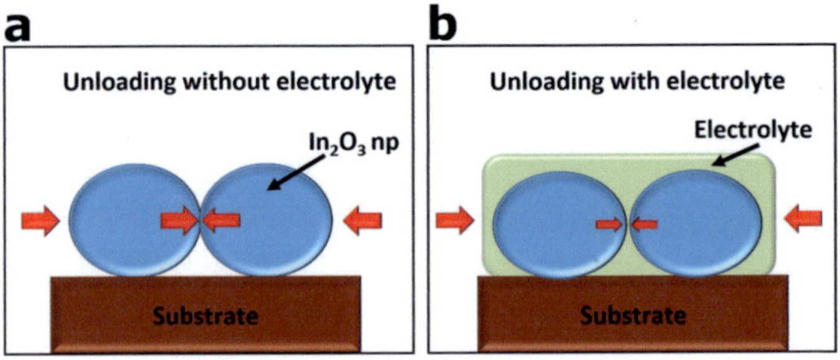

Figure 5.8: Schematic model to explain the lattice strain evolution of ink-jet printed In_2O_3 film on a polymer substrate with and without electrolyte encapsulation of the film during a) loading and b) unloading of the sample.

The penetration of the electrolyte into the nanoparticle network provides a compliant interlayer between the nanoparticles. This prevents the frictional compression between the particles while unloading resulting in a decrease in the accommodation of compressive strain within the In_2O_3 nanoparticles. In the study of the effect of multiple stacking of different materials including the np film, no significant difference in the lattice strain evolution of In_2O_3 np film deposited on bare PI and with an Au interlayer was observed. Further, the effect of electrolyte encapsulation is still observed. However, a significant strain drop is observed for the Au film. This can be explained as the dense and well bonded Au film forms channel cracks at a rather small tensile strain of about 1.5 %. Encapsulation does not have any influence in the lattice stain evolution of Au film as there is no penetration of the electrolyte, whereas the effect of film encapsulation is seen on the lattice strain evolution of the In_2O_3 np film due to the nanoporous nature of the film.

Good adhesion between the film and the substrate leads to a direct strain transfer from the substrate to the film, which is the case for the evaporated Au films. It reaches very high lattice strains and finally fractures at low strain as discussed above. For the Au electrodes the strain transfer is similar and a similar range in lattice strain is obtained, however, no significant strain drop is observed, and the lattice strain only slightly decreases during continuous straining of the sample. This can be explained because of the higher probability of presence of micro-cracks and other artefacts in a continuous film as compared to the patterned Au electrodes with much smaller volume and surface area as schematically shown in figure 5.9.

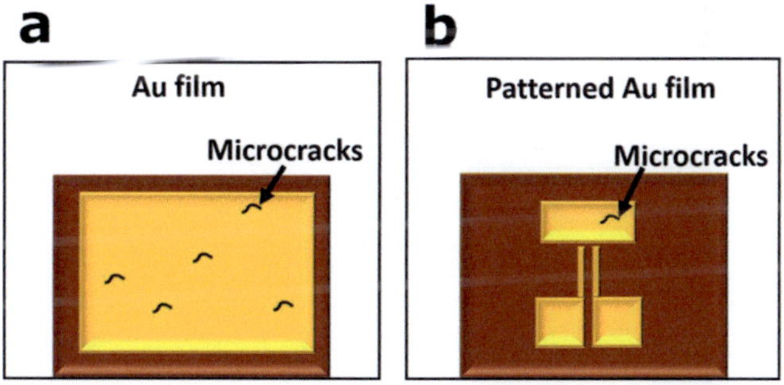

Figure 5.9: Schematic model to explain lower fracture probability in Au electrodes compared to a continuous Au film.

For the mechanical characterization of ink-jet printed In_2O_3 nanoparticulate films on polyamide substrate it is shown successfully that synchrotron based XRD characterization can be used to analyze the strain transfer in nanoparticulate films and small structures although the signal-to-noise ratio for patterned films is very low. It is also shown that the s-XRD technique can be used to analyze individual materials in a multi-material system during a single tensile test. The effect of material stacking as well as geometrical constraints on the lattice strain evolution is studied. The increase in lattice strain is higher for thicker In_2O_3 np films as compared to thinner samples. The instantaneous decrease in lattice strain of the In_2O_3 films while unloading is similar irrespective of the thickness of the printed indium oxide film.

This behavior can be explained by assuming the nanoparticulate film as a network of loosely connected particles. Due to the compression on the particle connections during unloading an additional force transfer from particle to particle contributes to the lattice strain evolution of the np films. The encapsulation by polymer electrolyte also influences the strain transfer during unloading as the encapsulation shields the lattice strain evolution by providing a compliant interlayer between the particles and hence reducing the effect of force transfer at the particle contact points. An introduction of an Au interlayer does not affect the lattice strain evolution of the printed In_2O_3 np films. The geometrical constraints of the transistor design and transistor orientation with respect to the loading direction do not have a significant effect on the lattice strain evolution of the incorporated materials.

The lattice strain range in Au is an order of magnitude higher than in the nanoparticulate In_2O_3 films. Due to the dense microstructure and good adhesion of Au, the strain transfer from the substrate to the Au film is high. For a nanoparticulate film, the nanoporous microstructure reduces the strain transfer from the substrate to the film by the structural compliance of the particle network.

However, the physical contact between the particles and the functional properties of the material are retained during straining up to about 8 % strain. It has been reported earlier by Baby et al. (Baby et al., 2015) that the In_2O_3 nanoparticulate ink which has been used for this study can be used to produce high quality thin film transistors on glass substrate at room temperature. This shows a very high potential of nanoparticulate films for application as flexible electronics.

6 BST-PMMA MIM capacitor

6.1 Characterization of flexible capacitor

The characterization and testing of nanocomposite Metal Insulator Metal Capacitor (MIM) on Polyethylene terephthalate (PET) substrate is discussed. The Barium strontium titanate/Poly methyl methacrylate (BST/PMMA) nanocomposite ink, which is used in this study, is developed by Morten Mikolajek in the group of Dr. Joachim Binder KIT IAM-KWT. It shows good printability with an Ohnesorge number of Oh = 0.2 as well as a very homogeneous drying behavior on Al_2O_3 substrates (Mikolajek et al., 2015). The BST dispersion used has a very small particle size distribution in the range of $d_{particles}$ = 60 – 130 nm. This leads to a very good long-term stability of the dispersion as well as a uniform particle distribution in the printed films.

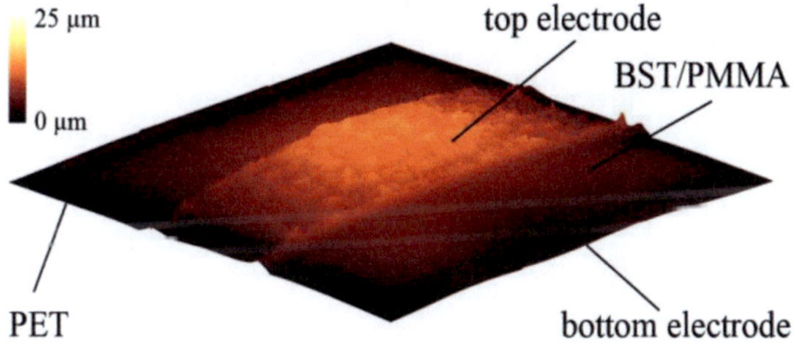

Figure 6.1: Topography of a fully ink-jet printed BST/PMMA MIM capacitor with Ag top and bottom electrodes on PET substrate.

For the preparation of flexible capacitors, the nanocomposite ink was printed in between two nanoparticulate Ag electrodes on a PET substrate (see chapter 3.1.3). The printed capacitors have an effective area of 4 mm^2. The topography of the capacitor area is exemplarily shown in figure 6.1. It can be seen that the BST/PMMA film shows a very smooth and homogeneous structure. This is essential for the fabrication of printed multi-layer components with smooth and strong interfaces.

To investigate the effect of annealing temperature on the mechanical and electrical performance of the Ag electrodes uniaxial tensile tests were performed on printed Ag electrodes annealed at different temperature and the corresponding change in resistance was recorded during the tensile tests. The comparison of the change in resistance of the Ag films is shown in figure 6.2. The initial film resistance of the printed Ag films decreases with increasing annealing temperature. However, the minimum strain that the film can withstand before losing its electrical conductivity by crack formation also decreases with increase in annealing temperature.

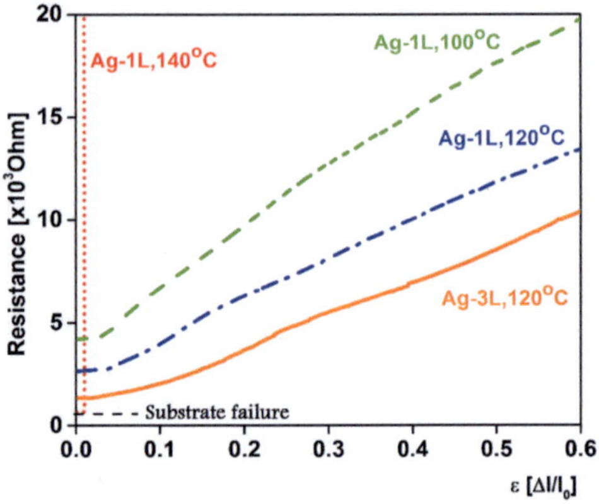

Figure 6.2: a) Change in resistance of ink-jet printed Ag films on PET substrate annealed at different temperatures. L denotes the number of printing passes.

Samples annealed at 140°C already failed below 1 % tensile strain. This complete failure is not due to fragmentation of the Ag electrode but due to the increase in the brittleness and sudden fracture of the PET substrate as the processing temperature approaches its glass transition temperature (Vicca et al., 2010). Based on these results it is decided to anneal all the printed samples at 120°C for further studies as the Ag films show an optimum balance of good electrical conductivity as well as stretchability. Figure 6.3a shows a cross-sectional SEM image of the all-printed capacitor.

The thickness of the top Ag layer, BST composite layer and the bottom layer are measured as 1.7 μm, 8.36 μm and 0.52 μm, respectively. The interfaces between the BST layer and the top and the bottom Ag electrodes are smooth and well defined. Figures 6.3b-e show details of the Ag top electrode and the BST layer at high and low magnification, respectively. It can be seen that, both, the printed Ag layer and the BST/PMMA nanocomposite film show a homogeneous nanoporous microstructure. Moreover, along with a homogeneous topography, the microstructure of the printed composite layer is very important for large area applications as the dielectric and the mechanical properties of the composite films strongly depend on the microstructure, which can be improved by a homogeneous dispersion of the nanoparticles in the polymer matrix.

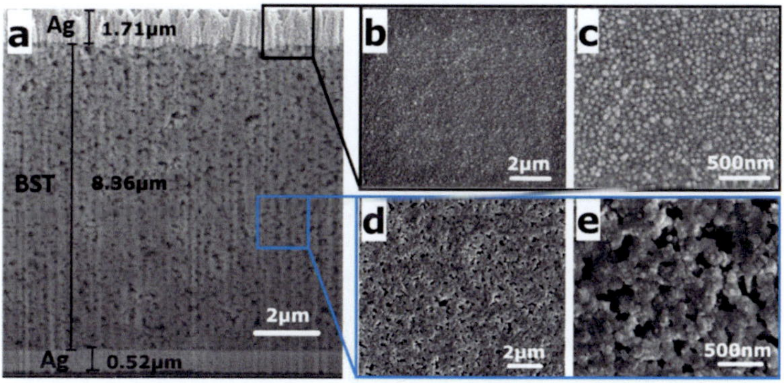

Figure 6.3: Microstructural characterization of printed MIM capacitors (a) cross-sectional SEM image of the fully printed MIM capacitor (b)-(c) SEM image of the Top Ag electrode and (d)-(e) the BST/PMMA layer at low and high magnification.

Therefore, a good compatibility of the ceramic and polymer phase is necessary. This is achieved using PMMA as the polymer matrix due to the good chemical interaction between the oxide ceramic surface and the ester side chains of PMMA. The dielectric film shows little porosity due to the shrinkage of PMMA during the temperature treatment above its glass transition temperature of $T_g \approx 105\,°C$. Besides, there is some inhomogeneity in the form of thin patches with inhomogeneous film thickness within the top Ag electrode as shown in figure 6.4, which occurs due to the printing process.

The printability of the Ag ink is ensured as the same printing parameters are used for both the top and bottom electrode, but the wetting behavior of the ink is different for the PET substrate and the BST/PMMA surface.

The capacitance for the as prepared devices measured at a frequency of 1 kHz and at a signal voltage of 1 V, lies in the range of 160-175 pF. The average value of the capacitance was measured as 165 pF based on three consecutive measurements.

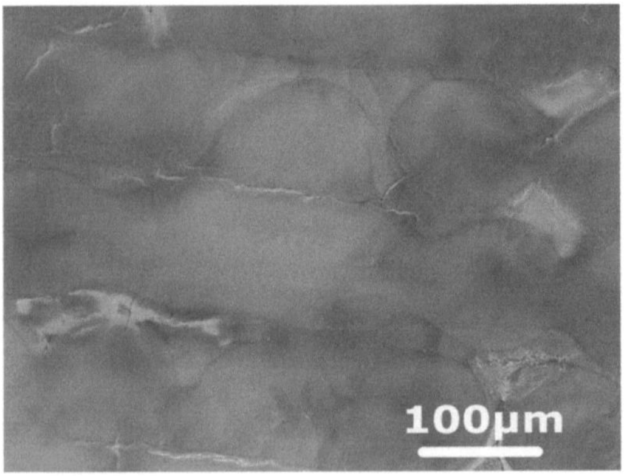

Figure 6.4: SEM image showing inhomogeneity in printed Ag top electrode.

Similarly, the Dissipation Factor (D) value was determined as 0.10 at 1 kHz. Both the measurements are shown in figure 6.5. The small scatter of the measured values for the capacitance further confirmed the homogeneity of the printed structures and the low D value (<1) further confirm a high-quality device fabrication.

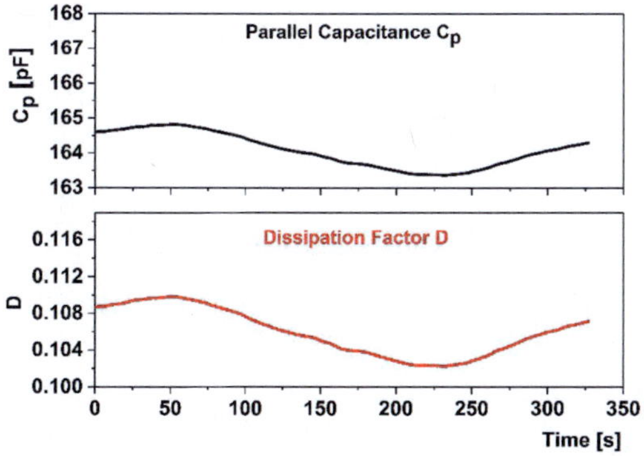

Figure 6.5: Measured capacitance and dissipation factor of a fully printed MIM capacitor on PET Substrate at 1kHz.

The theoretical capacitance of the capacitor was calculated to be 159.13 pF using equation 2.2. In this case, an effective capacitive area (A) of 4mm^2, a thickness of the dielectric layer (d) of 8.3 µm (calculated from figure 6.3a) and a relative permittivity for the BST/PMMA matrix (ε_r) of 37 at 1 kHz is used. The actual measured value of the capacitance of the printed capacitor is very close to the theoretically calculated capacitance at the same frequency. This confirms the good quality of the printed layers as well as the quality of the interfaces between the different layers of the printed capacitors.

6.2 Tensile testing of flexible capacitors

Uniaxial Tensile tests are conducted on the printed capacitors to understand the influence of external strain on the performance of the devices. The capacitance of the device is recorded during the mechanical test.

Figure 6.6 shows the evolution in measured capacitance of the printed capacitors and change in resistance of the printed top and bottom electrodes with respect to applied strain.

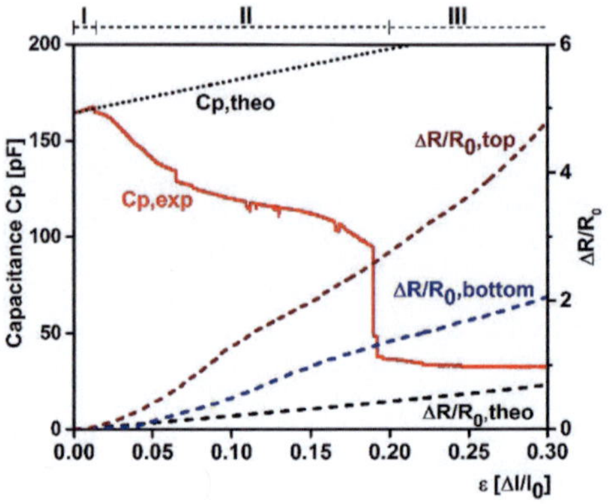

Figure 6.6: Evolution of capacitance $C_{p,exp}$, $\Delta R/R_{0,top}$ and $\Delta R/R_{0,bottom}$ resistance of the top and bottom Ag electrode, respectively, during the tensile test on fully printed MIMcapacitors on PET substrate. The theoretical change in capacitance $C_{p,theo}$ and resistance $\Delta R/R_{0,theo}$ are also shown for comparison.

The curves are an average representation of the electro-mechanical testing on 10 different samples having similar boundary conditions. The value of capacitance at first increases till 2 % tensile strain and then starts decreasing gradually until around 7-8 %, where a small drop is observed. Upon further straining another sharp drop in capacitance at 20 % of strain is observed. The dotted lines in figure 6.6 show the change in resistance ($\Delta R/R_0$) of the top Ag layer, bottom Ag layer and the theoretical line with respect to applied strain at a constant applied voltage of 1 V, where R_0 represents the initial resistance of the film at 1 V and ΔR represents change in resistance with the application of strain.

The theoretical line ($\Delta R/R_{0,theo}$) = $2\varepsilon + \varepsilon^2$ is calculated by considering the fact that the volume of active material is constant. ($\Delta R/R_{0,top}$) of the top Ag electrode starts deviating from the theoretical line at around 4 %

tensile strain, where the first few cracks in the Ag electrodes are observed, and continue to increase thereafter with increase in applied strain indicating further formation of cracks in the Ag film.

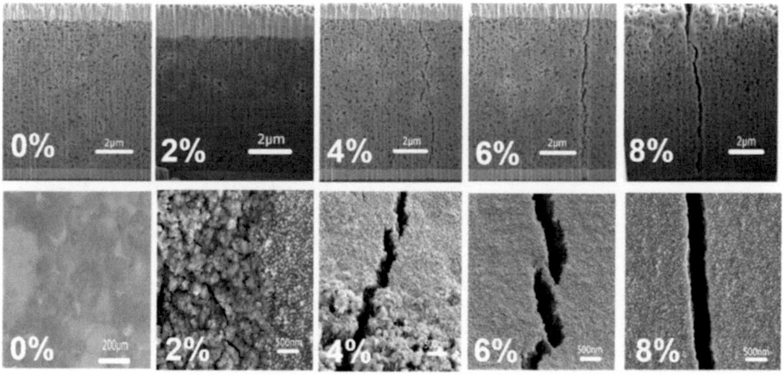

Figure 6.7: SEM images of top) cross sections, and bottom) top views of the fully printed capacitor after loading to 0 %, 2 %, 4 %, 6 % and 8 % tensile strain.

($\Delta R/R_{0,bottom}$) for the bottom Ag electrode starts deviating from the theoretical curve at around 7-8 % of tensile strain indicating the formation of through-thickness cracks within the capacitor. The theoretical line for the change in capacitance of the printed capacitor with respect to applied strain is calculated as $C_{p,theo} = C_0(1+ \varepsilon)$, where C_0 represents the initial capacitance of the capacitor at 1 V and 1k Hz. The corresponding assumption for this equation is increasing area due to decrease in film thickness due to Poisson contraction.

Planar and cross-sectional SEM images of the tested capacitors were taken, and the images are collected in figure 6.7. The upper row of pictures shows cross-sectional images of the capacitors and the bottom row of pictures shows planar views of the Ag top electrode. Cracks in the BST/PMMA layer are first observed at 2 % of tensile strain and in the Ag top electrodes cracks form at about 4 % of tensile strain. The crack spacing of the individual layers is determined by taking an average of the number of cracks along a line of 1000 µm and 10 such lines were analysed per sample. Figure 6.8 shows the average crack spacing for the BST/PMMA layer and the top Ag layer at a tensile train of 4 %, 6 %, and 8 %, respectively. The crack spacing in the BST/PMMA layer is lower as compared to top Ag layer and decreases significantly during straining.

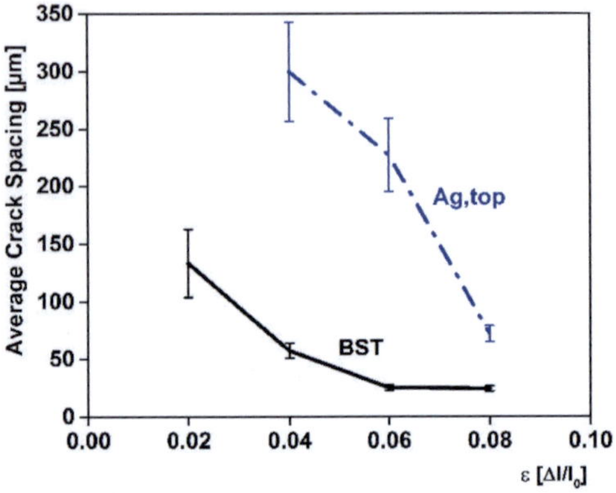

Figure 6.8: Average crack spacing within the Ag top electrode and the BST/PMMA layer as a function of tensile strain.

6.3 Discussion

It is observed that straining of the printed capacitors leads to initiation and propagation of cracks within the individual layers. This leads to a gradual decrease in capacitance. The cross-sectional images of the strained printed capacitors show the mechanisms of crack propagation within the printed capacitor and confirm that micro-cracks are initiated at individual spots within the BST/PMMA layer at around 2 % which subsequently propagate to the top electrode at around 4 % and the bottom electrodes at around 7-8 %, respectively. For the top view of the tested capacitors, up to 2 % of strain, hardly any crack is visible in the top Ag electrode. At around 4% of strain, cracks start forming in the thinner regions of the Ag electrode. The thickness of the Ag top electrode is inhomogeneous and elliptical region a with uneven height and thickness are observed. This is probably due to printing process as the structure comes from the individual droplets and insufficient wetting of the ink.

Beyond 6 % strain, crack opening increases and cracking spacing decreases further with increase in applied strain. At around 8 % strain, the cracks in the BST/PMMA layer are propagating into the bottom Ag electrode. Subsequently the crack spacing decreases in all individual layers. Finally, at around 20 % of strain, an individual crack propagates across the thickness and width of the top electrode and leads to almost complete failure of the capacitor.

The development of vertical cracks within the BST layer ideally should not affect the capacitance of the plate capacitor as it will result in an arrangement of parallel capacitors having electrical connection to each other, whose combined capacitance will be obtained by arithmetically adding them (Hayt, 1989) . However, as shown in figure 6.6, the capacitance of the printed capacitor measured at constant voltage and frequency continuously decreases with the application of strain. Therefore, the decrease in capacitance must result from additional failure modes. Based on the change in capacitance with respect to applied strain, three different zones may be defined (also see figure 6.6).

Zone I:

Up to ~2 % strain, the decrease in BST/PMMA film thickness due to Poisson contraction leads to a reduction of the distance between the Ag electrodes and leads to a small increase in capacitance. The slope of this increase in capacitance follows the theoretical line. At 2 % strain, first cracks appear in BST/PMMA layer.

Zone II:

Between 2 % and 20 % strain, the capacitance decreases continuously due to the combination of several effects. The schematics of these effects are sketched in figure 6.9. First, the decrease in capacitance is due to a loss in effective area as the thinnest regions of the Ag top electrode loose their conductivity. Before the application of strain, the Ag top electrodes, although locally very thin, is conducting everywhere. However, beyond a certain strain, the thinnest regions get non-conducting and hence yield to a decrease in active capacitor area. This phenomenon has been confirmed by individual local two-point measurements. At 4 % cracks form in the top Ag electrode and single cracks in the BST/PMMA layer also propagate into the top electrode. And at around 7-8 % first cracks in the BST/PMMA layer enter the bottom Ag electrode. All these mechanism lead to further fragmentation of the capacitor and reduction of capacitive area and further reduce the overall capacitance.

In addition, fringe or edge effects also come into consideration in this zone. Due to the crack formation in the top electrode, the capacitor is separated into several individual capacitors as a part of the capacitive area is separated from the total area and act as individual capacitor itself. On one hand the capacitance of the MIM capacitor in this case will decrease due to the formation of cracks. On the other hand, fringe effect increases the capacitance of the device due to additional electric field lines at the corner of the individual parallel plate capacitors. However, the intensity is very low as long as both the width and height of the capacitor are significantly larger than the plate separation (Hayt, 1989).

Zone III:

At around 20 % tensile strain a sudden decrease in capacitance is observed. At such high strain, there is propagation of a dominant crack through the thickness and width of the top electrode, which leads to a partial or a total electrical circuit cut-off leading to final failure of the device. Often delamination is observed in the vicinity of the crack.

Further indicators of capacitor degradation are the increasing relative resistances of the top and bottom Ag electrodes. Increasing deviation from the theoretical curve indicates ongoing crack formation and propagation within the film. Based on this criterion, degradation of the top and bottom electrodes starts at around 3 and 6 %, respectively. This corresponds nicely with the SEM observations.

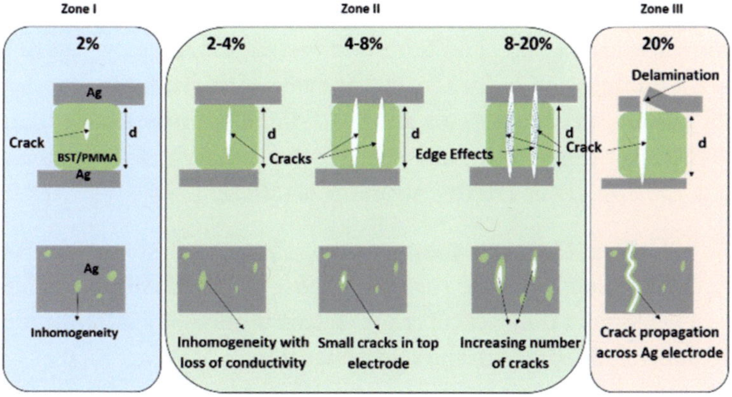

Figure 6.9: Schematics of possible failure mechanisms to understand the steady decrease of the capacitance of the fully printed MIM capacitor on PET with increasing strain.

The decrease in capacitance with increasing tensile strain is correlated with the increase in resistance of the Ag electrodes due to the formation and propagation of cracks. The lower value of crack spacing within BST layer as compared to the top Ag layer further confirms the argument that cracks are initiated within BST/PMMA layer and later propagate to the top and bottom electrodes of the printed capacitor.

The propagation of vertical cracks in the BST/PMMA layer ideally should not deteriorate the capacitance of the devices significantly. In contrast, the formation of through-thickness cracks in the Ag electrodes will lead to a partial deactivation of the capacitive area. Therefore, the decrease in the measured capacitance is primarily because of inhomogeneity and early fracture of the top Ag electrode. Therefore, one of the ways to further improve the mechanical performance of the capacitors is to increase the homogeneity and quality of the top Ag layer. As a proof of concept, a new ink (SunTronic™ Nanosilver EMD5730) is used to print the top electrode over the printed BST-PMMA layer while keeping all other fabrication parameters the same. For the bottom electrode the old Ag Ink (Silver Jet DGP 40LT-15C) is still used is to print the bottom electrode on the bare PET surface.

Uniaxial Tensile tests were performed on the optimized capacitors under the same environmental conditions and testing parameters. Figure 6.10 shows the comparison of the tensile behavior of the MIM capacitors with the old and the new Ag top electrode. It should be noted here that although there is no significant difference in the initial capacitance the evolution of the capacitance during straining is different.

For the optimized capacitor, the capacitance again at first increases up to 2 % strain, but subsequently remains almost constant with only slightly decreasing slope as the vertical cracks forms in the BST/PMMA layer got trapped at the interfaces to the electrodes which prevents further crack propagation into the Ag electrodes. Also, the increase in resistance of the new top Ag electrode $(\Delta R/R_0)_{Top\text{-}new}$ is lower than for the old one $(\Delta R/R_0)_{Top\text{-}old}$ confirming the latter and modulated onset of cracking in the optimized top electrode. At around 24 % strain a sharp drop in the measured capacitance is then observed due to the failure of the PET substrate, which defines the ultimate failure of the printed MIM capacitor.

In this study a successful fabrication route of fully printed BST/PMMA capacitors on compliant substrate with competitive performance at a low processing temperature of 120°C has been demonstrated. It also includes a comparative study about the effect of processing temperature on the mechanical performance of individual layers of the printed device. For the tensile tests it has been found that the printed capacitors are consistent with its performance for a strain of about 2 %, and then a gradual decrease is observed which corresponds to the crack formation in the BST/PMMA layer which subsequently propagates to the top and the bottom electrodes of the device leading to final failure at 20 % of tensile strain.

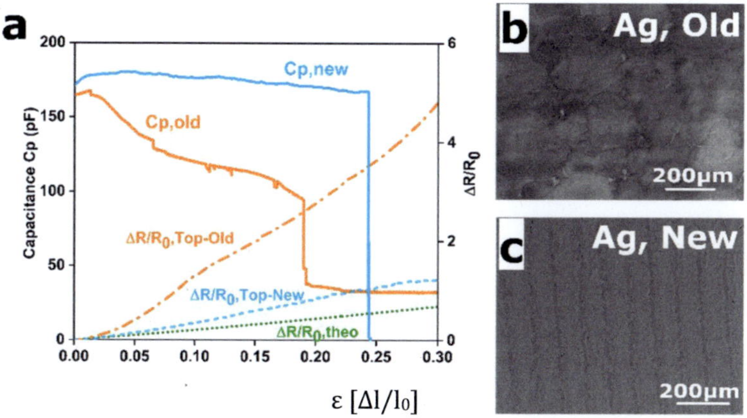

Figure 6.10: (a) Comparison of the uniaxial tensile test on fully printed MIM capacitors on PET printed with the old and with the new optimized Ag top electrode (b) & c) SEM images of the top electrodes. On the surface of new top electrode, a more homogeneous line pattern is observed as compared to irregular and hilly surface of the old Ag electrode.

It is shown that the morphology of the Ag top electrode plays a crucial role for the degradation of the capacitance. Therefore, optimization of the top Ag layer can significantly improve the strainability of printed MIM capacitors. The crack formation in the dielectric is not as important for the functionality of the printed capacitor but can be even further moderated by optimizing the ratio of the BST and the polymer in the dielectric nanocomposite.

7 Summary and outlook

Three different systems have been studied in order to understand the influence of strain on the performance and functionality of ink-jet printed thin films as well as printed devices on flexible substrate. Ink-jet printed In_2O_3 thin film transistors prepared through a precursor route on polyimide substrate, ink-jet printed In_2O_3 nanoparticulate films on polyimide substrate as well as, ink-jet printed BST/PMMA nanocomposite metal insulator metal capacitor on PET substrate have been investigated. For characterization of flexible printed films and devices different electro-mechanical testing techniques were established.

For printed In_2O_3 transistors, uniaxial tensile tests and bending fatigue tests were performed and transistor parameters were determined before and after the test to evaluate the transistor behavior under monotonic and cyclic loading, respectively. For nanoparticulate In_2O_3 films, in-situ tensile tests in combination with synchrotron X-ray diffraction were performed to evaluate the lattice strain evolution within nanoparticulate films and small structures. For printed BSST/PMMA capacitors, uniaxial tensile tests were conducted, and the device capacitance was monitored continuously during the test to evaluate the degradation in device capacitance with the onset of cracks.

Successful fabrication of a thin film transistor on a flexible substrate with very high electrical performance has been shown. The effect of mechanical strain on the electrical performance of the printed indium oxide transistor has been investigated. It has been found that the transistor fabricated on Kapton® E substrate through precursor route has a good electrical performance with a field-effect mobility of 70.61 cm^2/Vs and an I_{ON}/I_{OFF} ratio in the order of 10^7 at a very low applied voltage and has a reasonably good mechanical flexibility. The transistors were fully operational up to 10,000 cycles at an applied bending strain of 2 % and a bending radius of 1.25 mm under both tensile and compressive loading conditions. For uniaxial tension, the transistors were also fully operational up to 2 % strain and then start deteriorating due to formation of cracks within the oxide channel layer and due to delamination at the interface between the Au electrode and the indium oxide layer. The importance of the channel film quality in the overall reliability of the devices is also discussed. Fracture in In_2O_3 film is visible; however other interfaces are also important for reliability of printed In_2O_3 transistors.

In-situ synchrotron XRD testing was used for the mechanical characterization of ink-jet printed In_2O_3 nanoparticulate films as well as for structures having lateral dimensions. It small is shown that the synchrotron XRD technique yields lattice strain evolution of individual layers (e.g. In_2O_3 and Au) with in the transistor structure which can be analysed during a single tensile test. The influence of several process and design parameters on the lattice strain evolution has been investigated. During loading, maximum lattice strain increases for increasing film thickness. Lattice strain becomes instantaneously compressive while unloading irrespective of the thickness of the printed indium oxide film. This behavior can be explained by assuming the film as a network of loosely connected particles. Deposition of dielectric layer over the channel layer is a crucial process of transistor fabrication.

Therefore, to understand the effect of polymer electrolyte encapsulation on the lattice strain evolution in the nanoparticulate films, the In_2O_3 np films were encapsulated with solid polymer electrolyte and investigated. During the loading of the sample encapsulation does not affect the lattice strain evolution of the material, however, during unloading; the compliant interlayer moderates the lattice strain evolution. Introduction of Au interlayer does not affect the lattice strain evolution of the printed In_2O_3 films. For small structures, the signal to noise ratio becomes inferior but lattice strain could be still evaluated. Preparation of inorganic functional films through a nanoparticulate route is very promising for the development of flexible printed devices with high mechanical performance and reliability, as nanoparticulate films are more tolerant to crack formation as compared to precursor films. Along with excellent particle connectivity nanoparticulate films have low strain transfer from the substrate and thus sustain higher strains before failure.

Finally, a successful fabrication route for preparation of fully printed BST/PMMA capacitors on compliant substrate with competitive performance at a low processing temperature of 120°C has been developed. A comparative study about the effect of processing temperature on the mechanical performance of individual layers of the device is also shown. Uniaxial tensile tests show that the printed MIM capacitor are consistent with its performance for strains up to 2 %, and then a gradual decrease in capacitance is observed corresponding to crack formation in the BST/PMMA layer which subsequently propagates to the top and the bottom Ag electrodes of the device leading to final failure at 20 % of tensile strain.

It is evident that the drastic decrease in the capacitance value of the printed capacitors is primarily related to failure of the top Ag electrode. As a proof of concept, that a more uniform deposition of the top Ag electrode can significantly improve the stability of the printed capacitor, an optimised capacitor is shown which has a stable capacitance until the PET substrate fails at 25 % tensile strain. The occurrence of cracks in the dielectric can be further moderated by optimizing the ratio of the BST and polymer in the dielectric nanocomposite material.

List of Figures

List of Tables

References

Anderson, J. C. (1972). Applications of thin films in microelectronics. *Thin Solid Films, 12*(1), 1–15. https://doi.org/10.1016/0040-6090(72)90386-0

Anthony, J. E., Facchetti, A., Heeney, M., Marder, S. R., & Zhan, X. (2010). n-Type Organic Semiconductors in Organic Electronics. *Advanced Materials, 22*(34), 3876–3892. https://doi.org/10.1002/adma.200903628

Aoki, A., & Sasakura, H. (1970). Tin Oxide Thin Film Transistors. *Japanese Journal of Applied Physics, 9*(5), 582–582. https://doi.org/10.1143/JJAP.9.582

Arzt, E. (1998). Size effects in materials due to microstructural and dimensional constraints: a comparative review. *Acta Materialia, 46*(16), 5611–5626. https://doi.org/10.1016/S1359-6454(98)00231-6

Avis, C., Hwang, H. R., & Jang, J. (2014). Effect Of Channel Layer Thickness On The Performance Of Indium–Zinc–Tin Oxide Thin Film Transistors Manufactured By Inkjet Printing. *ACS Applied Materials & Interfaces, 6*(14), 10941–10945. https://doi.org/10.1021/am501153w

Baby, T. T., Garlapati, S. K., Dehm, S., Häming, M., Kruk, R., Hahn, H., & Dasgupta, S. (2015). A General Route toward Complete Room Temperature Processing of Printed and High Performance Oxide Electronics. *ACS Nano, 9*(3), 3075–3083. https://doi.org/10.1021/nn507326z

Baiano, A. (2009). *Single grain TFTs for high speed flexible electronics.* Delft University of Technology. Retrieved from http://www.narcis.nl/publication/RecordID/oai:tudelft.nl:uuid:ee432065-0bfe-467a-a100-8d8ce96659d7

Banger, K. K., Yamashita, Y., Mori, K., Peterson, R. L., Leedham, T., Rickard, J., & Sirringhaus, H. (2011). Low-temperature, high-performance solution-processed metal oxide thin-film transistors formed by a "sol–gel on chip" process. *Nature Materials, 10*(1), 45–50. https://doi.org/10.1038/nmat2914

Berggren, M., Nilsson, D., & Robinson, N. D. (2007). Organic materials for printed electronics. *Nature Materials, 6*(1), 3–5. https://doi.org/10.1038/nmat1817

Beuth, J. L. (1992). Cracking of thin bonded films in residual tension. *International Journal of Solids and Structures, 29*(13), 1657–1675. https://doi.org/10.1016/0020-7683(92)90015-L

Boesen, G. F., & Jacobs, J. E. (1968). ZnO field-effect transistor. *Proceedings of the IEEE, 56*(11), 2094–2095. https://doi.org/10.1109/PROC.1968.6813

Bragg, W. H., & Bragg, W. L. (1913). The Reflection of X-rays by Crystals. *Proceedings of the Royal Society A: Mathematical, Physical and Engineering Sciences, 88*(605), 428–438. https://doi.org/10.1098/rspa.1913.0040

Brenner, S. S. (1956). Tensile Strength of Whiskers. *Journal of Applied Physics, 27*(12), 1484–1491. https://doi.org/10.1063/1.1722294

Cantarella, G., Münzenrieder, N., Petti, L., Member, S., Vogt, C., Member, S., Member, S. (2015). Flexible In – Ga – Zn – O Thin-Film Transistors on Elastomeric Substrate Bent to 2 . 3 % Strain. *IEEE Electron Device Letters, 36*(8), 781–783. https://doi.org/10.1109/LED.2015.2442271

Chen, H., Rim, Y. S., Jiang, C., & Yang, Y. (2015). Low-Impurity High-Performance Solution-Processed Metal Oxide Semiconductors via a Facile Redox Reaction. *Chemistry of Materials, 27*(13), 4713–4718. https://doi.org/10.1021/acs.chemmater.5b01347

Chen, M. (2003). Deformation Twinning in Nanocrystalline Aluminum. *Science, 300*(5623), 1275–1277. https://doi.org/10.1126/science.1083727

Cherenack, K. H., Kattamis, A. Z., Hekmatshoar, B., Sturm, J. C., & Wagner, S. (2007). Amorphous-Silicon Thin-Film Transistors Fabricated at 300. *IEEE Electron Device Letters, 28*(11), 1004–1006. https://doi.org/10.1109/LED.2007.907411

Chiu, S. L., Leu, J., & Ho, P. S. (1994). Fracture of metal-polymer line structures. I. Semiflexible polyimide. *Journal of Applied Physics, 76*(9), 5136–5142. https://doi.org/10.1063/1.357227

Choi, C.-H., Han, S.-Y., Su, Y.-W., Fang, Z., Lin, L.-Y., Cheng, C.-C., & Chang, C. (2015). Fabrication of high-performance, low-temperature solution processed amorphous indium oxide thin-film transistors using a volatile nitrate precursor. *Journal of Materials Chemistry C*, *3*(4), 854–860. https://doi.org/10.1039/C4TC01568A

Choi, C.-H., Lin, L.-Y., Cheng, C.-C., & Chang, C. -h. (2015). Printed Oxide Thin Film Transistors: A Mini Review. *ECS Journal of Solid State Science and Technology*, *4*(4), P3044–P3051. https://doi.org/10.1149/2.0071504jss

Chopra, K. L., Paulson, P. D., & Dutta, V. (2004). Thin-film solar cells: an overview. *Progress in Photovoltaics: Research and Applications*, *12*(23), 69–92. https://doi.org/10.1002/pip.541

Chrisey, D. B. (2000). MATERIALS PROCESSING: The Power of Direct Writing. *Science*, *289*(5481), 879–881. https://doi.org/10.1126/science.289.5481.879

Cordill, M. J., Glushko, O., Kreith, J., Marx, V. M., & Kirchlechner, C. (2015). Measuring electro-mechanical properties of thin films on polymer substrates. *Microelectronic Engineering*, *137*, 96–100. https://doi.org/10.1016/j.mee.2014.08.002

Dasgupta, S., Gottschalk, S., Kruk, R., & Hahn, H. (2008). A nanoparticulate indium tin oxide field-effect transistor with solid electrolyte gating. *Nanotechnology*, *19*(43), 435203. https://doi.org/10.1088/0957-4484/19/43/435203

Dasgupta, S., Kruk, R., Mechau, N., & Hahn, H. (2011). Inkjet Printed, High Mobility Inorganic-Oxide Field Effect Transistors Processed at Room Temperature. *ACS Nano*, *5*(12), 9628–9638. https://doi.org/10.1021/nn202992v

Derby, B. (2010). Inkjet Printing of Functional and Structural Materials: Fluid Property Requirements, Feature Stability, and Resolution. *Annual Review of Materials Research*, *40*(1), 395–414. https://doi.org/10.1146/annurev-matsci-070909-104502

Doerner, M. F., Gardner, D. S., & Nix, W. D. (1986). Plastic properties of thin films on substrates as measured by submicron indentation hardness and substrate curvature techniques. *Journal of Materials Research*, *1*(6), 845–851. https://doi.org/10.1557/JMR.1986.0845

Eberl Christoph. (2010). Digital Image Correlation and Tracking - File Exchange - MATLAB Central. Retrieved April 26, 2017, https://de.mathworks.com/trackingmatlabcentral/fileexchange/12413-digital-image-correlation.

Erdem Alaca, B., Saif, M. T. A., & Sehitoglu, H. (2002). On the interface debond at the edge of a thin film on a thick substrate. *Acta Materialia*, *50*(5), 1197–1209. https://doi.org/10.1016/S1359-6454(01)00421-9

Evans, A. G., & Hutchinson, J. W. (1995). The thermomechanical intigrity of thin films and multilayers. *Acta Metallurgica Et Materialia*, *43*(7), 2507–2530. Retrieved from http://www.seas.harvard.edu/hutchinson/papers/416.pdf

Filipovic, L., & Selberherr, S. (2015). Performance and Stress Analysis of Metal Oxide Films for CMOS-Integrated Gas Sensors. *Sensors*, *15*(4), 7206–7227. https://doi.org/10.3390/s150407206

Flinn, P. A., Gardner, D. S., & Nix, W. D. (1987). Measurement and Interpretation of stress in aluminum-based metallization as a function of thermal history. *IEEE Transactions on Electron Devices*, *34*(3), 689–699. https://doi.org/10.1109/T-ED.1987.22981

Forrest, S. R. (2004). The path to ubiquitous and low-cost organic electronic appliances on plastic. *Nature*, *428*(6986), 911–918. https://doi.org/10.1038/nature02498

Frank, S., Handge, U. A., Olliges, S., & Spolenak, R. (2009). The relationship between thin film fragmentation and buckle formation: Synchrotron-based in situ studies and two-dimensional stress analysis. *Acta Materialia*, *57*(5), 1442–1453. https://doi.org/10.1016/j.actamat.2008.11.023

Friederich, A., Kohler, C., Nikfalazar, M., Wiens, A., Sazegar, M., Jakoby, R., Binder, J. R. (2014). Microstructure and microwave properties of inkjet printed barium strontium titanate thick-films for tunable microwave devices. *Journal of the European Ceramic Society, 34*(12), 2925–2932. https://doi.org/10.1016/j.jeurceramsoc.2014.04.007

Fukuda, K., Takeda, Y., Yoshimura, Y., Shiwaku, R., Tran, L. T., Sekine, T.,Tokito, S. (2014). Fully-printed high-performance organic thin-film transistors and circuitry on one-micron-thick polymer films. *Nature Communications, 5*(May), 4147. https://doi.org/10.1038/ncomms5147

Garlapati, S. K., Baby, T. T., Dehm, S., Hammad, M., Chakravadhanula, V. S. K., Kruk, R., ... Dasgupta, S. (2015). Ink-Jet Printed CMOS Electronics from Oxide Semiconductors. *Small, 11*(29), 3591–3596. https://doi.org/10.1002/smll.201403288

Garlapati, S. K., Mishra, N., Dehm, S., Hahn, R., Kruk, R., Hahn, H., & Dasgupta, S. (2013). Electrolyte-Gated, High Mobility Inorganic Oxide Transistors from Printed Metal Halides. *ACS Applied Materials & Interfaces, 5*(22), 11498–11502. https://doi.org/10.1021/am403131j

Gevorgian, S. (2009). *Ferroelectrics in Microwave Devices, Circuits and Systems*. London: Springer London. https://doi.org/10.1007/978-1-84882-507-9

Golas, P. L., Louie, S., Lowry, G. V., Matyjaszewski, K., & Tilton, R. D. (2010). Comparative Study of Polymeric Stabilizers for Magnetite Nanoparticles Using ATRP. *Langmuir, 26*(22), 16890–16900. https://doi.org/10.1021/la103098q

Griffin, A. J., Brotzen, F. R., & Dunn, C. F. (1992). Mechanical testing of thin metallic films. *Thin Solid Films, 220*, 265–270.

Grouchko, M., Kamyshny, A., Mihailescu, C. F., Anghel, D. F., & Magdassi, S. (2011). Conductive Inks with a "Built-In" Mechanism That Enables Sintering at Room Temperature. *ACS Nano, 5*(4), 3354–3359. https://doi.org/10.1021/nn2005848

Gruber, P. A., Arzt, E., & Spolenak, R. (2009). Brittle-to-ductile transition in ultrathin Ta/Cu film systems. *Journal of Materials Research, 24*(6), 1906–1918. https://doi.org/10.1557/jmr.2009.0252

Gruber, P. A., Böhm, J., Onuseit, F., Wanner, A., Spolenak, R., & Arzt, E. (2008). Size effects on yield strength and strain hardening for ultra-thin Cu films with and without passivation: A study by synchrotron and bulge test techniques. *Acta Materialia*, *56*(10), 2318–2335. https://doi.org/10.1016/j.actamat.2008.01.027

Gruber, P., Böhm, J., Wanner, A., Sauter, L., Spolenak, R., & Arzt, E. (2004). Size Effect on Crack Formation in Cu/Ta and Ta/Cu/Ta Thin Film Systems. *MRS Proceedings*, *821*, P2.7. https://doi.org/10.1557/PROC-821-P2.7

Gudmundson, P., & Wikström, A. (2002). Stresses in thin films and interconnect lines. *Microelectronic Engineering*, *60*(1–2), 17–29. https://doi.org/10.1016/S0167-9317(01)00577-9

Hall, E. O. (1951). The Deformation and Ageing of Mild Steel: III Discussion of Results. *Proc. Phys. Soc. B*, *64*, 747–53. Retrieved from http://iopscience.iop.org/0370-1301/64/9/303

Hayt, W. H. (1989). *Engineering electromagnetics* (5th ed.). McGraw-Hill, New York.Retrieved from hhttps://trove.nla.gov.au/work/6713212

Herlogsson, L., Coelle, M., Tierney, S., Crispin, X., & Berggren, M. (2010). Low-Voltage Ring Oscillators Based on Polyelectrolyte-Gated Polymer Thin-Film Transistors. *Advanced Materials*, *22*(1), 72–76. https://doi.org/10.1002/adma.200901850

Hommel, M., Kraft, O., & Arzt, E. (1999). A new method to study cyclic deformation of thin films in tension and compression. *Journal of Materials Research*, *14*(6), 2373–2376. https://doi.org/10.1557/JMR.1999.0317

Huang, M., Suo, Z., & Ma, Q. (2002). Plastic ratcheting induced cracks in thin film structures. *Journal of the Mechanics and Physics of Solids*, *50*(5), 1079–1098. https://doi.org/10.1016/S0022-5096(01)00113-2

Hutchinson, J. W., & Suo, Z. (1991). Mixed Mode Cracking in Layered Materials. In *Advances in Applied Mechanics* (Vol. 29, pp. 63–191). https://doi.org/10.1016/S0065-2156(08)70164-9

Hwan Hwang, Y., Seo, J.-S., Moon Yun, J., Park, H., Yang, S., Ko Park, S.-H., & Bae, B.-S. (2013). An "aqueous route" for the fabrication of low-temperature-processable oxide flexible transparent thin-film transistors on plastic substrates. *NPG Asia Materials*, *5*(4), e45. https://doi.org/10.1038/am.2013.11

Jang, D., Kim, D., & Moon, J. (2009). Influence of fluid physical properties on ink-jet printability. *Langmuir*, *25*(5), 2629–2635. https://doi.org/10.1021/la900059m

Jeong, S., Lee, J. Y., Ham, M. H., Song, K., Moon, J., Seo, Y. H., ... Choi, Y. (2013). Bendable thin-film transistors based on sol-gel derived amorphous Ga-doped In2O3 semiconductors. *Superlattices and Microstructures*, *59*, 21–28. https://doi.org/10.1016/j.spmi.2013.03.026

Jeong, S., & Moon, J. (2012). Low-temperature, solution-processed metal oxide thin film transistors. *Journal of Materials Chemistry*, *22*(4), 1243–1250. https://doi.org/10.1039/C1JM14452A

Jiang, H., Khang, D.-Y., Song, J., Sun, Y., Huang, Y., & Rogers, J. A. (2007). Finite deformation mechanics in buckled thin films on compliant supports. *Proceedings of the National Academy of Sciences*, *104*(40), 15607–15612. https://doi.org/10.1073/pnas.0702927104

Jiang, H., Sun, Y., Rogers, J. A., & Huang, Y. (2008). Post-buckling analysis for the precisely controlled buckling of thin film encapsulated by elastomeric substrates. *International Journal of Solids and Structures*, *45*(7–8), 2014–2023. https://doi.org/10.1016/j.ijsolstr.2007.11.007

Ju, S., Facchetti, A., Xuan, Y., Liu, J., Ishikawa, F., Ye, P., ... Janes, D. B. (2007). Fabrication of fully transparent nanowire transistors for transparent and flexible electronics. *Nature Nanotechnology*, *2*(6), 378–384. https://doi.org/10.1038/nnano.2007.151

Jun, J. H., Park, B., Cho, K., & Kim, S. (2009). Flexible TFTs based on solution-processed ZnO nanoparticles. *Nanotechnology*, *20*(50), 505201. https://doi.org/10.1088/0957-4484/20/50/505201

Kang, Y. S., Ho, P. S., Knipe, R., & Tregilgas, J. (1996). Film Thickness Effect on Tensile Properties and Microstructures of Submicron Aluminum Thin Films on Polyimide. *MRS Proceedings*, *436*, 35. https://doi.org/10.1557/PROC-436-35

Keller, R.-M., Baker, S. P., & Arzt, E. (1998). Quantitative analysis of strengthening mechanisms in thin Cu films: Effects of film thickness, grain size, and passivation. *Journal of Materials Research, 13*(5), 1307–1317. https://doi.org/10.1557/JMR.1998.0186

Khang, D.-Y. (2006). A Stretchable Form of Single-Crystal Silicon for High-Performance Electronics on Rubber Substrates. *Science, 311*(5758), 208–212. https://doi.org/10.1126/science.1121401

Kim, B.-J., Shin, H.-A.-S., Jung, S.-Y., Cho, Y., Kraft, O., Choi, I.-S., & Joo, Y.-C. (2013). Crack nucleation during mechanical fatigue in thin metal films on flexible substrates. *Acta Materialia, 61*(9), 3473–3481. https://doi.org/10.1016/j.actamat.2013.02.041

Kim, B., Haas, T., Friederich, A., Lee, J., Nam, D.-H., Binder, J. R., ... Kraft, O. (2014). Improving mechanical fatigue resistance by optimizing the nanoporous structure of inkjet-printed Ag electrodes for flexible devices. *Nanotechnology, 25*(12), 125706. https://doi.org/10.1088/0957-4484/25/12/125706

Kim, D.-H., Xiao, J., Song, J., Huang, Y., & Rogers, J. A. (2010). Stretchable, Curvilinear Electronics Based on Inorganic Materials. *Advanced Materials, 22*(19), 2108–2124. https://doi.org/10.1002/adma.200902927

Kim, D. H., & Rogers, J. A. (2008). Stretchable electronics: Materials strategies and devices. *Advanced Materials, 20*(24), 4887–4892. https://doi.org/10.1002/adma.200801788

Kim, H. S., Byrne, P. D., Facchetti, A., & Marks, T. J. (2008). High Performance Solution-Processed Indium Oxide Thin-Film Transistors. *Journal of the American Chemical Society, 130*(38), 12580–12581. https://doi.org/10.1021/ja804262z

Kim, M.-G., Kanatzidis, M. G., Facchetti, A., & Marks, T. J. (2011). Low-temperature fabrication of high-performance metal oxide thin-film electronics via combustion processing. *Nature Materials, 10*(5), 382–388. https://doi.org/10.1038/nmat3011

Kim, S., Won, S., Sim, G.-D., Park, I., & Lee, S. (2013). Tensile characteristics of metal nanoparticle films on flexible polymer substrates for printed electronics applications. *Nanotechnology, 24*(8), 85701. https://doi.org/10.1088/0957-4484/24/8/085701

Kim, Y.-H., Heo, J.-S., Kim, T.-H., Park, S., Yoon, M.-H., Kim, J., ... Park, S. K. (2012). Flexible metal-oxide devices made by room-temperature photochemical activation of sol–gel films. *Nature, 489*(7414), 128–132. https://doi.org/10.1038/nature11434

Kirstein, T. (2013). The future of smart-textiles development: new enabling technologies, commercialization and market trends. In *Multidisciplinary Know-How for Smart-Textiles Developers* (pp. 1–25). Elsevier. https://doi.org/10.1533/9780857093530.1

Klasens, H. A., & Koelmans, H. (1964). A tin oxide field-effect transistor. *Solid-State Electronics, 7*(9), 701–702. https://doi.org/10.1016/0038-1101(64)90057-7

Ko, S. H., Pan, H., Grigoropoulos, C. P., Luscombe, C. K., Fréchet, J. M. J., & Poulikakos, D. (2007). All-inkjet-printed flexible electronics fabrication on a polymer substrate by low-temperature high-resolution selective laser sintering of metal nanoparticles. *Nanotechnology, 18*(34), 345202. https://doi.org/10.1088/0957-4484/18/34/345202

Kohler, C., Friederich, A., Sazegar, M., Nikfalazar, M., Stemme, F., Wang, D., ... Binder, J. R. (2013). Effects of ZnO-B 2 O 3 Addition on the Micro-structure and Microwave Properties of Low-Temperature Sintered Barium Strontium Titanate (BST) Thick Films. *International Journal of Applied Ceramic Technology, 10*(s1), E200–E209. https://doi.org/10.1111/ijac.12116

Kraft, O., Gruber, P. a., Mönig, R., & Weygand, D. (2010). Plasticity in Confined Dimensions. *Annual Review of Materials Research, 40*(1), 293–317. https://doi.org/10.1146/annurev-matsci-082908-145409

Kraft, O., Hommel, M., & Arzt, E. (2000). X-ray diffraction as a tool to study the mechanical behaviour of thin films. *Materials Science and Engineering: A, 288*(2), 209–216. https://doi.org/10.1016/S0921-5093(00)00876-5

Kraft, O., Schwaiger, R., & Wellner, P. (2001). Fatigue in thin films: lifetime and damage formation. *Materials Science and Engineering: A, 319–321*, 919–923. https://doi.org/10.1016/S0921-5093(01)00990-X

Kraft, O., & Volkert, C. A. (2001). Mechanical Testing of Thin Films and Small Structures. *Advanced Engineering Materials, 3*(3), 99–110. https://doi.org/10.1002/1527-2648(200103)3:3<99::AID-ADEM99>3.0.CO;2-2

Kwon, J.-Y., Lee, D.-J., & Kim, K.-B. (2011). Review paper: Transparent amorphous oxide semiconductor thin film transistor. *Electronic Materials Letters, 7*(1), 1–11. https://doi.org/10.1007/s13391-011-0301-x

Lacour, S. P., Chan, D., Wagner, S., Li, T., & Suo, Z. (n.d.). Mechanisms of reversible stretchability of thin metal films on elastomeric substrates. https://doi.org/10.1063/1.2201874

Lee, C. Y., Lin, M. Y., Wu, W. H., Wang, J. Y., Chou, Y., Su, W. F., ... Lin, C. F. (2010). Flexible ZnO transparent thin-film transistors by a solution-based process at various solution concentrations. *Semiconductor Science and Technology, 25*(10), 105008. https://doi.org/10.1088/0268-1242/25/10/105008

Lee, H.-Y., Yi, S.-M., Lee, J.-H., Lee, H.-S., Hyun, S., & Joo, Y.-C. (2010). Effects of bending fatigue on the electrical resistance in metallic films on flexible substrates. *Metals and Materials International, 16*(6), 947–951. https://doi.org/10.1007/s12540-010-1213-2

Lee, J.-H., Kim, N.-R., Kim, B.-J., & Joo, Y.-C. (2012). Improved mechanical performance of solution-processed MWCNT/Ag nanoparticle composite films with oxygen-pressure-controlled annealing. *Carbon, 50*(1), 98–106. https://doi.org/10.1016/j.carbon.2011.07.057

Lee, S. J., Ko, J., Nam, K.-H., Kim, T., Lee, S. H., Kim, J. H., ... Myoung, J.-M. (2016). Fully Solution-Processed and Foldable Metal-Oxide Thin-Film Transistor. *ACS Applied Materials & Interfaces*, acsami.6b00950. https://doi.org/10.1021/acsami.6b00950

Li, T., Huang, Z. Y., Xi, Z. C., Lacour, S. P., Wagner, S., & Suo, Z. (2005). Delocalizing strain in a thin metal film on a polymer substrate. *Mechanics of Materials, 37*(2–3 SPEC. ISS.), 261–273. https://doi.org/10.1016/j.mechmat.2004.02.002

Li, T., & Suo, Z. (2006). Deformability of thin metal films on elastomer substrates. *International Journal of Solids and Structures, 43*(7–8), 2351–2363. https://doi.org/10.1016/j.ijsolstr.2005.04.034

Lilienfeld Julius Edgar. (1930). Method and apparatus for controlling electric currents. US. Retrieved from http://www.google.de/patents/US1745175

Lim, J., Kim, J. J. J., Yoon, Y. J., Kim, H., Yoon, H. G., Lee, S.-N. N., & Kim, J. J. J. (2012). All-inkjet-printed Metal-Insulator-Metal (MIM) capacitor. *Current Applied Physics, 12*(SUPPL. 1), 14–17. https://doi.org/10.1016/j.cap.2011.04.035

Lin, Y.-H., Faber, H., Zhao, K., Wang, Q., Amassian, A., McLachlan, M., & Anthopoulos, T. D. (2013). High-Performance ZnO Transistors Processed Via an Aqueous Carbon-Free Metal Oxide Precursor Route at Temperatures Between 80-180 °C. *Advanced Materials, 25*(31), 4340–4346. https://doi.org/10.1002/adma.201301622

Liu, Y., Cui, T., & Varahramyan, K. (2003). All-polymer capacitor fabricated with inkjet printing technique. *Solid-State Electronics, 47*(9), 1543–1548. https://doi.org/10.1016/S0038-1101(03)00082-0

Lohmiller, J. A. (2013). Investigation of deformation mechanisms in nanocrystalline metals and alloys by in situ synchrotron X-ray diffraction.

Lohmiller, J., Baumbusch, R., Kerber, M. B., Castrup, A., Hahn, H., Schafler, E.,Gruber, P. A. (2013). Following the deformation behavior of nanocrystalline Pd films on polyimide substrates using in situ synchrotron XRD. *Mechanics of Materials, 67*, 65–73. https://doi.org/10.1016/j.mechmat.2013.04.010

Lohmiller, J., Baumbusch, R., Kraft, O., & Gruber, P. A. (2013). Differentiation of Deformation Modes in Nanocrystalline Pd Films Inferred from Peak Asymmetry Evolution Using In Situ X-Ray Diffraction. *Physical Review Letters, 110*(6), 66101. https://doi.org/10.1103/PhysRevLett.110.066101

Lohmiller, J., Grewer, M., Braun, C., Kobler, A., Kübel, C., Schüler, K., … Gruber, P. A. (2014). Untangling dislocation and grain boundary mediated plasticity in nanocrystalline nickel. *Acta Materialia, 65*(12), 295–307. https://doi.org/10.1016/j.actamat.2013.10.071

Lohmiller, J., Kobler, A., Spolenak, R., & Gruber, P. A. (2013). The effect of solute segregation on strain localization in nanocrystalline thin films: Dislocation glide vs. grain-boundary mediated plasticity. *Applied Physics Letters, 102*(24), 241916. https://doi.org/10.1063/1.4811743

Lu, N., Wang, X., Suo, Z., & Vlassak, J. (2007). Metal films on polymer substrates stretched beyond 50 %. *Applied Physics Letters, 91*(22), 221909. https://doi.org/10.1063/1.2817234

Macionczyk, F., & Brückner, W. (1999). Tensile testing of AlCu thin films on polyimide foils. *Journal of Applied Physics, 86*(9), 4922–4929. https://doi.org/10.1063/1.371461

Magdassi, S. (2009). *The Chemistry of Inkjet Inks*. World scientific. https://doi.org/10.1142/6869

Mahajan, A., Francis, L. F., & Frisbie, C. D. (2014). Facile Method for Fabricating Flexible Substrates with Embedded, Printed Silver Lines. *ACS Applied Materials & Interfaces, 6*(2), 1306–1312. https://doi.org/10.1021/am405314s

Mativenga, M., Choi, M. H., Choi, J. W., & Jang, J. (2011). Transparent flexible circuits based on amorphous-indium-gallium-zincoxide thin-film transistors. *IEEE Electron Device Letters, 32*(2), 170–172. https://doi.org/10.1109/LED.2010.2093504

Matthews, J. (1974). Defects in epitaxial multilayers I. Misfit dislocations. *Journal of Crystal Growth, 27*, 118–125. https://doi.org/10.1016/0022-0248(74)90424-2

Melinex ® polyester film. (2012). Retrieved November 21, 2017, fromhttp://www.koenigkunststoffe.de/produkte/melinex/melinex-r-st506.pdf

Meng, C., Liu, C., Chen, L., Hu, C., & Fan, S. (2010). Highly flexible and all-solid-state paperlike polymer supercapacitors. *Nano Letters, 10*(10), 4025–4031. https://doi.org/10.1021/nl1019672

Mikolajek, M., Friederich, A., Kohler, C., Rosen, M., Rathjen, A., Krüger, K., & Binder, J. R. (2015). Direct Inkjet Printing of Dielectric Ceramic/Polymer Composite Thick Films. *Advanced Engineering Materials, 17*(9), 1294–1301. https://doi.org/10.1002/adem.201400451

Minami, T. (2005). Transparent conducting oxide semiconductors for transparent electrodes. *Semiconductor Science and Technology, 20*(4), S35–S44. https://doi.org/10.1088/0268-1242/20/4/004

Minemawari, H., Yamada, T., Matsui, H., Tsutsumi, J., Haas, S., Chiba, R., ... Hasegawa, T. (2011). Inkjet printing of single-crystal films. *Nature, 475*(7356), 364–367. https://doi.org/10.1038/nature10313

Münzenrieder, N., Cantarella, G., Vogt, C., Petti, L., Büthe, L., Salvatore, G. A., ... Tröster, G. (2015). Stretchable and Conformable Oxide Thin-Film Electronics. *Advanced Electronic Materials, 1*(3), n/a-n/a. https://doi.org/10.1002/aelm.201400038

Nickel, F., Haas, T., Wegner, E., Bahro, D., Salehin, S., Kraft, O., ... Colsmann, A. (2014). Mechanically robust, ITO-free, 4.8 % efficient, all-solution processed organic solar cells on flexible PET foil. *Solar Energy Materials and Solar Cells, 130*(February), 317–321. https://doi.org/10.1016/j.solmat.2014.07.005

Nix, W. D. (1989). Mechanical properties of thin films. *Metallurgical Transactions A, 20*(11), 2217–2245. https://doi.org/10.1007/BF02666659

Noyan, I. C., & Sheikh, G. (1993). X-ray tensile testing of thin films. *Journal of Materials Research, 8*(4), 764–770. https://doi.org/10.1557/JMR.1993.0764

Oliver, W. C., & Pharr, G. M. (1992). An improved technique for determining hardness and elastic modulus using load and displacement sensing indentation experiments. *Journal of Materials Research, 7*(6), 1564–1583. https://doi.org/10.1557/JMR.1992.1564

Olliges, S., Gruber, P. A., Orso, S., Auzelyte, V., Ekinci, Y., Solak, H. H., & Spolenak, R. (2008). In situ observation of cracks in gold nano-interconnects on flexible substrates. *Scripta Materialia, 58*(3), 175–178. https://doi.org/10.1016/j.scriptamat.2007.09.037

Park, J. H., Oh, J. Y., Han, S. W., Lee, T. Il, & Baik, H. K. (2015). Low-temperature, solution-processed ZrO2:B Thin film: A bifunctional inorganic/organic interfacial glue for flexible thin-film transistors. *ACS Applied Materials and Interfaces, 7*(8), 4494–4503. https://doi.org/10.1021/acsami.5b00036

Park, J. H., Yoo, Y. B., Lee, K. H., Jang, W. S., Oh, J. Y., Chae, S. S., ... Baik, H. K. (2013). Boron-Doped Peroxo-Zirconium Oxide Dielectric for High-Performance, Low-Temperature, Solution-Processed Indium Oxide Thin-Film Transistor. *ACS Applied Materials & Interfaces, 5*(16), 8067–8075. https://doi.org/10.1021/am402153g

Park, M.-J., Yun, D.-J., Ryu, M.-K., Yang, J.-H., Pi, J.-E., Kwon, O.-S., Yoon, S.-M. (2015). Improvements in the bending performance and bias stability of flexible InGaZnO thin film transistors and optimum barrier structures for plastic poly(ethylene naphthalate) substrates. *J. Mater. Chem. C, 3*(18), 4779–4786. https://doi.org/10.1039/C5TC00048C

Park, S.-I., Ahn, J.-H., Feng, X., Wang, S., Huang, Y., & Rogers, J. A. (2008). Theoretical and Experimental Studies of Bending of Inorganic Electronic Materials on Plastic Substrates. *Advanced Functional Materials, 18*(18), 2673–2684. https://doi.org/10.1002/adfm.200800306

Petch N J. (1953). The cleavage strength of polycrystals. *J. Iron Steel Inst, 174*, 25–8. http://garfield.library.upenn.edu/classics1982/A1982NM35100001.pdf

Petti, L., Frutiger, A., M??nzenrieder, N., Salvatore, G. A., B??the, L., Vogt, C., ... Tr??ster, G. (2015). Flexible quasi-vertical In-Ga-Zn-O thin-film transistor with 300-nm channel length. *IEEE Electron Device Letters, 36*(5), 475–477. https://doi.org/10.1109/LED.2015.2418295

Petti, L., Münzenrieder, N., Vogt, C., Faber, H., Büthe, L., Cantarella, G., ... Tröster, G. (2016). Metal oxide semiconductor thin-film transistors for flexible electronics. *Applied Physics Reviews, 3*(2), 21303. https://doi.org/10.1063/1.4953034

Quintero, A. V., Molina-Lopez, F., Smits, E. C. P., Danesh, E., van den Brand, J., Persaud, K., ... Briand, D. (2016). Smart RFID label with a printed multisensor platform for environmental monitoring. *Flexible and Printed Electronics, 1*(2), 25003. https://doi.org/10.1088/2058-8585/1/2/025003

Renault, P.-O., Badawi, K. F., Bimbault, L., Goudeau, P., Elkaïm, E., & Lauriat, J. P. (1998). Poisson's ratio measurement in tungsten thin films combining an x-ray diffractometer with in situ tensile tester. *Applied Physics Letters, 73*(14), 1952–1954. https://doi.org/10.1063/1.122332

Rim, Y. S., Chen, H., Liu, Y., Bae, S. H., Kim, H. J., & Yang, Y. (2014). Direct light pattern integration of low-temperature solution-processed all-oxide flexible electronics. *ACS Nano*, *8*(9), 9680–9686. https://doi.org/10.1021/nn504420r

Sanchez, J. E., & Arzt, E. (1992). Effects of grain orientation on hillock formation and grain growth in aluminum films on silicon substrates. *Scripta Metallurgica et Materialia*, *27*(3), 285–290. https://doi.org/10.1016/0956-716X(92)90513-E

Schmitt, B., Brönnimann, C., Eikenberry, E. ., Gozzo, F., Hörmann, C., Horisberger, R., & Patterson, B. (2003). Mythen detector system. *Nuclear Instruments and Methods in Physics Research Section A: Accelerators, Spectrometers, Detectors and Associated Equipment*, *501*(1), 267–272. https://doi.org/10.1016/S0168-9002(02)02045-4

Sekitani, T., Iba, S., Kato, Y., Noguchi, Y., Someya, T., & Sakurai, T. (2005). Ultraflexible organic field-effect transistors embedded at a neutral strain position. *Applied Physics Letters*, *87*(17), 173502. https://doi.org/10.1063/1.2115075

Sekitani, T., Zschieschang, U., Klauk, H., & Someya, T. (2010). Flexible organic transistors and circuits with extreme bending stability. *Nature Materials*, *9*(12), 1015–1022. https://doi.org/10.1038/nmat2896

Seo, J.-S., Jeon, J.-H., Hwang, Y. H., Park, H., Ryu, M., Park, S.-H. K., & Bae, B.-S. (2013). Solution-Processed Flexible Fluorine-doped Indium Zinc Oxide Thin-Film Transistors Fabricated on Plastic Film at Low Temperature. *Scientific Reports*, *3*(1), 1–9. https://doi.org/10.1038/srep02085

Sim, G.-D., Won, S., & Lee, S.-B. (2012). Tensile and fatigue behaviors of printed Ag thin films on flexible substrates. *Applied Physics Letters*, *101*(19), 191907. https://doi.org/10.1063/1.4766447

Singh, M., Haverinen, H. M., Dhagat, P., & Jabbour, G. E. (2010). Inkjet printing-process and its applications. *Advanced Materials*, *22*(6), 673–685. https://doi.org/10.1002/adma.200901141

Sirringhaus, H., Kawase, T., Friend, R. H., Shimoda, T., Inbasekaran, M., Wu, W., & Woo, E. P. (2000). High-resolution inkjet printing of all-polymer transistor circuits . *Science, 290*(December), 2123–2126. https://doi.org/10.1126/science.290.5499.2123

Song, K., Jeong, Y., Jun, T., Koo, C. Y., Kim, D., Woo, K., ... Moon, J. (2010). Low-temperature solution-deposited oxide thin-film transistors based on solution-processed organic-inorganic hybrid dielectrics. *Japanese Journal of Applied Physics, 49*(5 PART 2), 3–6. https://doi.org/10.1143/JJAP.49.05EB02

Song, K., Kim, D., Li, X., Jun, T., Moon, J., Jeong, Y., & Moon, J. (2009). Solution processed invisible all-oxide thin film transistors. *Journal of Materials Chemistry, 19*(46), 8881–8886. https://doi.org/10.1039/b912554j

Song, K., Noh, J., Jun, T., Jung, Y., Kang, H. Y., & Moon, J. (2010). Fully flexible solution-deposited ZnO thin-film transistors. *Advanced Materials, 22*(38), 4308–4312. https://doi.org/10.1002/adma.201002163

Sottos, N. R., Kandula, S., Geubelle, P. H., & Mechanics, A. (2005). Crack initiation in ultra thin patterned films. *Proceedings of the 11th International Conference on Fracture.*

Spaepen, F., & Shull, A. L. (1996). Mechanical properties of thin films and multilayers. *Current Opinion in Solid State and Materials Science, 1*(5), 679–683. https://doi.org/10.1016/S1359-0286(96)80051-9

Sriprachuabwong, C., Srichan, C., Lomas, T., & Tuantranont, a. (2010). Simple RC low pass filter circuit fabricated by unmodified desktop inkjet printer. *Electrical Engineering/Electronics Computer Telecommunications and Information Technology (ECTI-CON), 2010 International Conference on*, 3–6.

Suo, Z., Ma, E. Y., Gleskova, H., & Wagner, S. (1999). Mechanics of rollable and foldable film-on-foil electronics. *Applied Physics Letters, 74*(8), 1177–1179. https://doi.org/10.1063/1.123478

Suo, Z., Vlassak, J., & Wagner, S. (2005). Micromechanics of macroelectronics. *China Particuology, 3*(6), 321–328. https://doi.org/10.1016/S1672-2515(07)60210-3

Sven Olliges a, Patric A. Gruber b, Vaida Auzelyte c, Yasin Ekinci c, Harun H. Solak c, R. S. (2007). Tensile strength of gold nanointerconnects without the influence of strain gradients. *Acta Materialia*, *55*(15), 5201–5210. https://doi.org/10.1016/j.actamat.2007.05.039

Takahashi, T., Takei, K., Adabi, E., Fan, Z., Niknejad, A. M., & Javey, A. (2010). Parallel array InAs nanowire transistors for mechanically bendable, ultrahigh frequency electronics. *ACS Nano*, *4*(10), 5855–5860. https://doi.org/10.1021/nn1018329

Thiemann, S., Sachnov, S. J., Pettersson, F., Bollström, R., Österbacka, R., Wasserscheid, P., & Zaumseil, J. (2014). Cellulose-Based Ionogels for Paper Electronics. *Advanced Functional Materials*, *24*(5), 625–634. https://doi.org/10.1002/adfm.201302026

Tickle, A. C. (1969). *Thin-film Transistors; a New Approach to Microelectronics - Andrew C. Tickle - Google Books*. Wiley. https://books.google.de/books?id=X9M6vgAACAAJ&dq=inauthor:%22Andrew+C.+Tickle%22&hl=en&sa=X&ved=0ahUKEwiewOD_8ubQAhVqIsAKHQlqAbwQ6AEIHDAA

Toraya, H. (1990). Array-type universal profile function for powder pattern fitting. *Journal of Applied Crystallography*, *23*(6), 485–491. https://doi.org/10.1107/S002188989000704X

Vescio, G., López-Vidrier, J., Leghrib, R., Cornet, A., & Cirera, A. (2016). Flexible inkjet printed high-k HfO 2 -based MIM capacitors. *Journal of Materials Chemistry C*, *4*(9), 1804–1812. https://doi.org/10.1039/C5TC03307A

Vicca, P., Steudel, S., Smout, S., Raats, A., Genoe, J., & Heremans, P. (2010). A low-temperature-cross-linked poly(4-vinylphenol) gate-dielectric for organic thin film transistors. *Thin Solid Films*, *519*(1), 391–393. https://doi.org/10.1016/j.tsf.2010.08.009

Wagner, S., Lacour, S. P., Jones, J., Hsu, P. H. I. H. I., Sturm, J. C., Li, T., & Suo, Z. (2004). Electronic skin: Architecture and components. *Physica E: Low-Dimensional Systems and Nanostructures*, *25*(2–3 SPEC.ISS.), 326–334. https://doi.org/10.1016/j.physe.2004.06.032

Weimer, P. (1962). The TFT A New Thin-Film Transistor. *Proceedings of the IRE, 50*(6), 1462–1469. https://doi.org/10.1109/JRPROC.1962.288190

Woo, N. C., Cherenack, K., Tröster, G., & Spolenak, R. (2010). Designing micro-patterned Ti films that survive up to 10% applied tensile strain. *Applied Physics A: Materials Science and Processing, 100*(1), 281–285. https://doi.org/10.1007/s00339-010-5806-x

Wyss, A., Sologubenko, A. S., Mishra, N., Gruber, P. A., & Spolenak, R. (2017). Monitoring of stress–strain evolution in thin films by reflection anisotropy spectroscopy and synchrotron X-ray diffraction. *Journal of Materials Science, 52*(11), 6741–6753. https://doi.org/10.1007/s10853-017-0909-9

Xiang, Y., Li, T., Suo, Z., & Vlassak, J. J. (2005). High ductility of a metal film adherent on a polymer substrate. *Applied Physics Letters, 87*(16), 161910. https://doi.org/10.1063/1.2108110

Yeo, W. H., Kim, Y. S., Lee, J., Ameen, A., Shi, L., Li, M., ... Rogers, J. A. (2013). Multifunctional epidermal electronics printed directly onto the skin. *Advanced Materials, 25*(20), 2773–2778. https://doi.org/10.1002/adma.201204426

Yin, Z., Huang, Y., Bu, N., Wang, X., & Xiong, Y. (2010). Inkjet printing for flexible electronics: Materials, processes and equipments. *Chinese Science Bulletin, 55*(30), 3383–3407. https://doi.org/10.1007/s11434-010-3251-y

Yu, H. (2001). Edge effects in thin film delamination. *Acta Materialia, 49*(1), 93–107. https://doi.org/10.1016/S1359-6454(00)00293-7

Yu, X., Zeng, L., Zhou, N., Guo, P., Shi, F., Buchholz, D. B., ... Facchetti, A. (2015). Ultra-flexible, “invisible” thin-film transistors enabled by amorphous metal oxide/polymer channel layer blends. *Advanced Materials, 27*(14), 2390–2399. https://doi.org/10.1002/adma.201405400

Zhang, T., Hou, Z., Johnson, R. W., Del Castillo, L., Moussessian, A., Greenwell, R., & Blalock, B. J. (2009). Flexible electronics: Thin silicon die on flexible substrates. *IEEE Transactions on Electronics Packaging Manufacturing, 32*(4), 291–300. https://doi.org/10.1109/TEPM.2009.2028880

Zhu, L., & Wang, Q. (2012). Novel Ferroelectric Polymers for High Energy Density and Low Loss Dielectrics. *Macromolecules, 45*(7), 2937–2954. https://doi.org/10.1021/ma2024057

Zhu, Z. T., Menard, E., Hurley, K., Nuzzo, R. G., & Rogers, J. A. (2005). Spin on dopants for high-performance single-crystal silicon transistors on flexible plastic substrates. *Applied Physics Letters, 86*(13), 1–3. https://doi.org/10.1063/1.1894611

Zysset, C., Munzenrieder, N., Kinkeldei, T., Cherenack, K., & Troster, G. (2012). Woven active-matrix display. *IEEE Transactions on Electron Devices, 59*(3), 721–728. https://doi.org/10.1109/TED.2011.2180724

Schriftenreihe
des Instituts für Angewandte Materialien

ISSN 2192-9963

Band 1 Prachai Norajitra
Divertor Development for a Future Fusion Power Plant. 2011
ISBN 978-3-86644-738-7

Band 2 Jürgen Prokop
Entwicklung von Spritzgießsonderverfahren zur Herstellung von Mikrobauteilen durch galvanische Replikation. 2011
ISBN 978-3-86644-755-4

Band 3 Theo Fett
New contributions to R-curves and bridging stresses – Applications of weight functions. 2012
ISBN 978-3-86644-836-0

Band 4 Jérôme Acker
Einfluss des Alkali/Niob-Verhältnisses und der Kupferdotierung auf das Sinterverhalten, die Strukturbildung und die Mikrostruktur von bleifreier Piezokeramik $(K_{0,5}Na_{0,5})NbO_3$. 2012
ISBN 978-3-86644-867-4

Band 5 Holger Schwaab
Nichtlineare Modellierung von Ferroelektrika unter Berücksichtigung der elektrischen Leitfähigkeit. 2012
ISBN 978-3-86644-869-8

Band 6 Christian Dethloff
Modeling of Helium Bubble Nucleation and Growth in Neutron Irradiated RAFM Steels. 2012
ISBN 978-3-86644-901-5

Band 7 Jens Reiser
Duktilisierung von Wolfram. Synthese, Analyse und Charakterisierung von Wolframlaminaten aus Wolframfolie. 2012
ISBN 978-3-86644-902-2

Band 8 Andreas Sedlmayr
Experimental Investigations of Deformation Pathways in Nanowires. 2012
ISBN 978-3-86644-905-3

Die Bände sind unter www.ksp.kit.edu als PDF frei verfügbar oder als Druckausgabe bestellbar.

Band 9 Matthias Friedrich Funk
Microstructural stability of nanostructured fcc metals during cyclic deformation and fatigue. 2012
ISBN 978-3-86644-918-3

Band 10 Maximilian Schwenk
Entwicklung und Validierung eines numerischen Simulationsmodells zur Beschreibung der induktiven Ein- und Zweifrequenzrandschichthärtung am Beispiel von vergütetem 42CrMo4. 2012
ISBN 978-3-86644-929-9

Band 11 Matthias Merzkirch
Verformungs- und Schädigungsverhalten der verbundstranggepressten, federstahldrahtverstärkten Aluminiumlegierung EN AW-6082. 2012
ISBN 978-3-86644-933-6

Band 12 Thilo Hammers
Wärmebehandlung und Recken von verbundstranggepressten Luftfahrtprofilen. 2013
ISBN 978-3-86644-947-3

Band 13 Jochen Lohmiller
Investigation of deformation mechanisms in nanocrystalline metals and alloys by in situ synchrotron X-ray diffraction. 2013
ISBN 978-3-86644-962-6

Band 14 Simone Schreijäg
Microstructure and Mechanical Behavior of Deep Drawing DC04 Steel at Different Length Scales. 2013
ISBN 978-3-86644-967-1

Band 15 Zhiming Chen
Modelling the plastic deformation of iron. 2013
ISBN 978-3-86644-968-8

Band 16 Abdullah Fatih Çetinel
Oberflächendefektausheilung und Festigkeitssteigerung von niederdruckspritzgegossenen Mikrobiegebalken aus Zirkoniumdioxid. 2013
ISBN 978-3-86644-976-3

Band 17 Thomas Weber
Entwicklung und Optimierung von gradierten Wolfram/EUROFER97-Verbindungen für Divertorkomponenten. 2013
ISBN 978-3-86644-993-0

Band 18 Melanie Senn
Optimale Prozessführung mit merkmalsbasierter Zustandsverfolgung. 2013
ISBN 978-3-7315-0004-9

Band 19 Christian Mennerich
Phase-field modeling of multi-domain evolution in ferromagnetic shape memory alloys and of polycrystalline thin film growth. 2013
ISBN 978-3-7315-0009-4

Band 20 Spyridon Korres
On-Line Topographic Measurements of Lubricated Metallic Sliding Surfaces. 2013
ISBN 978-3-7315-0017-9

Band 21 Abhik Narayan Choudhury
Quantitative phase-field model for phase transformations in multi-component alloys. 2013
ISBN 978-3-7315-0020-9

Band 22 Oliver Ulrich
Isothermes und thermisch-mechanisches Ermüdungsverhalten von Verbundwerkstoffen mit Durchdringungsgefüge (Preform-MMCs). 2013
ISBN 978-3-7315-0024-7

Band 23 Sofie Burger
High Cycle Fatigue of Al and Cu Thin Films by a Novel High-Throughput Method. 2013
ISBN 978-3-7315-0025-4

Band 24 Michael Teutsch
Entwicklung von elektrochemisch abgeschiedenem LIGA-Ni-Al für Hochtemperatur-MEMS-Anwendungen. 2013
ISBN 978-3-7315-0026-1

Band 25 Wolfgang Rheinheimer
Zur Grenzflächenanisotropie von $SrTiO_3$. 2013
ISBN 978-3-7315-0027-8

Band 26 Ying Chen
Deformation Behavior of Thin Metallic Wires under Tensile and Torsional Loadings. 2013
ISBN 978-3-7315-0049-0

Die Bände sind unter www.ksp.kit.edu als PDF frei verfügbar oder als Druckausgabe bestellbar.

Band 27 Sascha Haller
Gestaltfindung: Untersuchungen zur Kraftkegelmethode. 2013
ISBN 978-3-7315-0050-6

Band 28 Nicht erschienen

Band 29 Gunnar Picht
Einfluss der Korngröße auf ferroelektrische Eigenschaften dotierter $Pb(Zr_{1-x}Ti_x)O_3$ Materialien. 2013
ISBN 978-3-7315-0106-0

Band 30 Esther Held
Eigenspannungsanalyse an Schichtverbunden mittels inkrementeller Bohrlochmethode. 2013
ISBN 978-3-7315-0127-5

Band 31 Pei He
On the structure-property correlation and the evolution of Nanofeatures in 12-13.5% Cr oxide dispersion strengthened ferritic steels. 2014
ISBN 978-3-7315-0141-1

Band 32 Jan Hoffmann
Ferritische ODS-Stähle – Herstellung, Umformung und Strukturanalyse. 2014
ISBN 978-3-7315-0157-2

Band 33 Wiebke Sittel
Entwicklung und Optimierung des Diffusionsschweißens von ODS Legierungen. 2014
ISBN 978-3-7315-0182-4

Band 34 Osama Khalil
Isothermes Kurzzeitermüdungsverhalten der hoch-warmfesten Aluminium-Knetlegierung 2618A (AlCu2Mg1,5Ni). 2014
ISBN 978-3-7315-0208-1

Band 35 Nicht erschienen

Band 36 Christoph Hage
Grundlegende Aspekte des 2K-Metallpulverspritzgießens. 2014
ISBN 978-3-7315-0217-3

Band 37 Bartłomiej Albiński
Instrumentierte Eindringprüfung bei Hochtemperatur für die Charakterisierung bestrahlter Materialien. 2014
ISBN 978-3-7315-0221-0

Band 38 Tim Feser
Untersuchungen zum Einlaufverhalten binärer alpha-Messinglegierungen unter Ölschmierung in Abhängigkeit des Zinkgehaltes. 2014
ISBN 978-3-7315-0224-1

Band 39 Jörg Ettrich
Fluid Flow and Heat Transfer in Cellular Solids. 2014
ISBN 978-3-7315-0241-8

Band 40 Melanie Syha
Microstructure evolution in strontium titanate Investigated by means of grain growth simulations and x-ray diffraction contrast tomography experiments. 2014
ISBN 978-3-7315-0242-5

Band 41 Thomas Haas
Mechanische Zuverlässigkeit von gedruckten und gasförmig abgeschiedenen Schichten auf flexiblem Substrat. 2014
ISBN 978-3-7315-0250-0

Band 42 Aron Kneer
Numerische Untersuchung des Wärmeübertragungsverhaltens in unterschiedlichen porösen Medien. 2014
ISBN 978-3-7315-0252-4

Band 43 Manuel Feuchter
Investigations on Joule heating applications by multiphysical continuum simulations in nanoscale systems. 2014
ISBN 978-3-7315-0261-6

Band 44 Alexander Vondrous
Grain growth behavior and efficient large scale simulations of recrystallization with the phase-field method. 2014
ISBN 978-3-7315-0280-7

Band 45 Tobias Kennerknecht
Fatigue of Micro Molded Materials – Aluminum Bronze and Yttria Stabilized Zirconia. 2014
ISBN 978-3-7315-0293-7

Band 46 Christopher Scherr
Elektrochemisches Verhalten von Lithium-Schwefel-Zellen mit unterschiedlicher Kathodenstruktur. 2015
ISBN 978-3-7315-0296-8

Band 47 Konstantin Frölich
Der Decal-Prozess zur Herstellung katalysatorbeschichteter Membranen für PEM-Brennstoffzellen. 2015
ISBN 978-3-7315-0334-7

Band 48 Benedikt Haspel
Werkstoffanalytische Betrachtung der Eigenschaften von mittels neuartiger RTM-Fertigungsprozesse hergestellten glasfaserverstärkten Polymerverbunden. 2015
ISBN 978-3-7315-0337-8

Band 49 Marco Berghoff
Skalenübergreifende Modellierung und Optimierung vom atomistischen kristallinen Phasenfeldmodell bis zur mesoskopischen Phasenfeldmethode. 2015
ISBN 978-3-7315-0416-0

Band 50 Michael Selzer
Mechanische und Strömungsmechanische Topologieoptimierung mit der Phasenfeldmethode. 2016
ISBN 978-3-7315-0431-3

Band 51 Michael Mahler
Entwicklung einer Auswertemethode für bruchmechanische Versuche an kleinen Proben auf der Basis eines Kohäsivzonenmodells. 2016
ISBN 978-3-7315-0441-2

Band 52 Christoph Bohnert
Numerische Untersuchung des Verformungs- und Bruchverhaltens von einkristallinem Wolfram auf mikroskopischer Ebene. 2016
ISBN 978-3-7315-0444-3

Band 53 Stefan Guth
Schädigung und Lebensdauer von Nickelbasislegierungen unter thermisch-mechanischer Ermüdungsbeanspruchung bei verschiedenen Phasenlagen. 2016
ISBN 978-3-7315-0445-0

Band 54 Markus Klinsmann
The Effects of Internal Stress and Lithium Transport on Fracture in Storage Materials in Lithium-Ion Batteries. 2016
ISBN 978-3-7315-0455-9

Die Bände sind unter www.ksp.kit.edu als PDF frei verfügbar oder als Druckausgabe bestellbar.

Band 55 Thomas Straub
Experimental Investigation of Crack Initiation in Face-Centered Cubic Materials in the High and Very High Cycle Fatigue Regime. 2016
ISBN 978-3-7315-0471-9

Band 56 Maren Lepple
Kupfer- und Eisenoxide als Konversions-Elektrodenmaterialien für Lithium-Ionen-Batterien: Thermodynamische und Elektrochemische Untersuchungen. 2016
ISBN 978-3-7315-0482-5

Band 57 Stefan Andreas Slaby
Charakterisierung und Bewertung der Zug- und Ermüdungseigenschaften von Mikrobauteilen aus 17-4PH Edelstahl. Ein Vergleich von mikropulverspritzgegossenem und konventionell hergestelltem Material. 2017
ISBN 978-3-7315-0484-9

Band 58 Kumar Ankit
Phase-field modeling of microstructural pattern formation in alloys and geological veins. 2016
ISBN 978-3-7315-0491-7

Band 59 Kuo Zhang
Characterization and Modeling of the Ratcheting Behavior of the Ferritic-Martensitic Steel P91. 2017
ISBN 978-3-7315-0503-7

Band 60 Nicht erschienen

Band 61 Fabian Lemke
Untersuchung des Sinterverhaltens von $SrTiO_3$ unter Berücksichtigung der Defektchemie. 2016
ISBN 978-3-7315-0510-5

Band 62 Johannes Kümmel
Detaillierte Analyse der Aufbauschneidenbildung bei der Trockenzerspanung von Stahl C45E mit Berücksichtigung des Werkzeugverschleißes. 2016
ISBN 978-3-7315-0518-1

Band 63 László Hagymási
Modellierung der Stoffübertragung beim Niederdruckcarbonitrieren mit Ammoniak und Acetylen. 2016
ISBN 978-3-7315-0568-6

Band 64 Reza Eslami
A novel micro-mechanical model for prediction of multiaxial high cycle fatigue at small scales. 2017
ISBN 978-3-7315-0583-9

Band 65 Sebastian Schulz
Phase-field simulations of multi-component solidification and coarsening based on thermodynamic datasets. 2017
ISBN 978-3-7315-0618-8

Band 66 Markus Stricker
Die Übertragung von mikrostrukturellen Eigenschaften aus der diskreten Versetzungsdynamik in Kontinuumsbeschreibungen. 2017
ISBN 978-3-7315-0658-4

Band 67 Luis Straßberger
Untersuchung und Modellierung des viskoplastischen Verformungsverhaltens oxidpartikelverstärkter Stähle. 2018
ISBN 978-3-7315-0674-4

Band 68 Mark Wobrock
Microplasticity of idealized single crystalline Ag cantilevers characterized with methods of high resolution. 2017
ISBN 978-3-7315-0682-9

Band 69 Amritesh Kumar
Micromechanical study on the deformation behaviour of directionally solidified NiAl-Cr eutectic composites. 2017
ISBN 978-3-7315-0694-2

Band 70 Johannes Hötzer
Massiv-parallele und großskalige Phasenfeldsimulationen zur Untersuchung der Mikrostrukturentwicklung. 2017
ISBN 978-3-7315-0693-5

Band 71 Thomas Hupfer
Herstellung von LATP für den Einsatz als Festkörperelektrolyt und dessen Eigenschaften. 2017
ISBN 978-3-7315-0702-4

Band 72 Florentin Pottmeyer
Schädigungsverhalten von in CFK-Laminaten eingebetteten Inserts unter bauteilnahen Beanspruchungen. 2017
ISBN 978-3-7315-0719-2

Band 73 Andres Höweling
Untersuchung der Hochvoltstabilität und Tiefentladung von dotierten $LiNi_{0,5}Mn_{1,5}O_4$-Hochvoltspinellen. 2018
ISBN 978-3-7315-0728-4

Band 74 Tabea Gisela Schwark
Deformation and Fracture Properties of the Soft Magnetic Composite Somaloy 700 3P on Different Length Scales. 2018
ISBN 978-3-7315-0759-8

Band 75 Klaudia Lichtenberg
Metallmatrixverbunde mit Verstärkungselementen aus metallischem Glas $Ni_{60}Nb_{20}Ta_{20}$ – Herstellung und Charakterisierung. 2018
ISBN 978-3-7315-0782-6

Band 76 Claudio Findeisen
Charakterisierung und Modellierung von instabilen Metamaterialien. 2019
ISBN 978-3-7315-0869-4

Band 77 Nilesha Mishra
Influence of strain on the functionality of ink-jet printed thin films and devices on flexible substrates. 2019
ISBN 978-3-7315-0853-3